Also by Nilofer Merchant

11 Rules for Creating Value in the #SocialEra

The New How

THE
POWER
OF
ONLYNESS

Make Your Wild Ideas Mighty
Enough to Dent the World

NILOFER MERCHANT

VIKING

VIKING

An imprint of Penguin Random House LLC
375 Hudson Street
New York, New York 10014
penguin.com

Illustrations on pages 37, 38, 43, 54, 69, 74, 96, 101, 107, 112, 130, 135, 141, 161, 166, 191, 196,
200, 218, 219, 220, 226, 232 by Chioma Ebinama

Library of Congress Cataloging-in-Publication Data

Names: Merchant, Nilofer, author.
Title: The power of onlyness : make your wild ideas mighty enough to dent the
world / Nilofer Merchant.
Description: New York : Viking, [2017] | Includes bibliographical references and index. |
Identifiers: LCCN 2017025380 (print) | LCCN 2017051942 (ebook) |
ISBN 9780698196155 (ebook) | ISBN 9780525429135 (hardcover)
Subjects: LCSH: Helping behavior.
Classification: LCC HM1146 (ebook) | LCC HM1146 .M47 2017 (print) | DDC 303.4—dc23
LC record available at https://lccn.loc.gov/2017025380

Printed in the United States of America
3   5   7   9   10   8   6   4

Set in Dante MT Std
Designed by Cassandra Garruzzo

Only that day dawns to which we are awake.

—Henry David Thoreau

For you, because you refuse to limit your ideas.

CONTENTS

THE POWER OF ONLYNESS

CHAPTER 1

Arriving at the Question

Tell me, what is it you plan to do with your
one wild and precious life?
—MARY OLIVER

VALUABLE IDEAS

Three teenage boys get the Boy Scouts of America to change its discriminatory policies. An older brother trying to save his sick younger sibling makes the entire health care industry address previously "incurable" diseases. Complete strangers come together to obtain justice for a seventy-year-old war crime.

The young, the sick, the neglected—these are not typically the people whose ideas are heard. Most often, whether ideas are considered or dismissed is based on who contributes them, and how powerful their sponsors—not the ideas themselves—are. So if the young, sick, and neglected can succeed in making a dent, what does their achievement mean for the rest of us—those of us who are told that our ideas can't be heard because our voices are too shrill, or because we lack certain credentials, or simply because the idea we're proposing is "too much"? Couldn't our ideas have a chance, too?

And don't they need to?

When I first began to write about this concept, in 2011, I struggled to describe it. I didn't want to argue only that new ideas and perspectives mattered to the modern creative economies—which they do—or that people gathered in networks could now scale projects that once only large hierarchical organizations could manage, or that changing times meant that we no longer had to "fit in" to organizations as a way to get things done. What I was seeking to propose, rather, was that anyone's—quite possibly everyone's—ideas mattered.

The key concept was that every one of the 7.5 billion humans on this planet has value to offer. How? You're standing in a spot in the world that *only* you stand in, a function of your history and experiences, visions, and hopes. From this spot where *only* you stand, you offer a distinct point of view, novel insights, and even groundbreaking ideas. Now that you can grow and realize those ideas through the power of networks, you have a new lever to move the world.

I tried to use existing vocabulary to express this concept, but nothing sufficed. The word I was searching for had to be a noun, one that would convey how value creation can come from anyone, and how even wild ideas now have a chance to flourish because networks allow anyone to bypass the standard gatekeepers and the frameworks they hold as true. As Sarah Green Carmichael—my editor at *Harvard Business Review*, for whom I was writing an article—and I twiddled back and forth, we finally realized no standard term worked.* Yet, because

---

* "Talent" was disqualified because it is typically a measure of labor that is credential dependent. For example, when someone has a degree, or they've already done the job and therefore have relevant experience, or their test scores are high enough to show aptitude, they are viewed as "talented." Thus, "talent" as it is often used fails to describe what I was addressing in the new ways of creating value, which is that anyone can contribute. In onlyness, there is always inherent capacity, with or without credentials.

Another word that seemed close to onlyness was "uniqueness," but the reason it can't serve as well as "onlyness" is because it is often contextual. You can be, for example, the "unique" person in the room if you are the twenty-five-year-old in a room full of

both of us had been the "only one" at different stages of our careers, moments when the dominant culture told us our particular "only" made our ideas marginal rather than meaningful, we thought, *Let's invert that. Let's reclaim the idea of each person's "only" as a strength.* With that flash of insight, the term "onlyness"* was born. Through the power of onlyness, an individual conceives an idea born of his narrative, nurtures it with the help of a community that embraces it, and, through shared action, makes the idea powerful enough to dent the world.

The storm that created that particular lightning bolt had in fact been brewing for some time. I had often wondered whether anyone could actually be eligible to have a shot at success, or whether people had to fit a particular profile to have their ideas be valued. The tension between those two alternatives had profound effects in my own life, so let me share three stories of how I discovered my own onlyness.

EIGHT BOOKS AND TWO OUTFITS

I didn't know my life would never be the same when, one day in 1986, I walked through my suburban Cupertino, California, neighborhood and into the local Winchell's Donut House. I thought I would be back at home in an hour, maybe two—four hours at most. I turned out to be deeply wrong.

Earlier that day, at age eighteen, I had come home to an unexpectedly full house of aunties making the fragrant, buttery rice and chicken dish Indians call *biryani*. It was impossible to identify amongst the medley of voices which one of them announced it to me first, but my

---

over-fifty-year-olds. But given that 50.5 percent of the world's 7.5 billion people are under thirty, youth is not that "unique," so the word lacks sufficient distinction. Onlyness says each of us 7.5 billion humans has the capacity to contribute, something *only* they have to offer.

* While a Google search returned no specific uses of this term at the time, I later learned "onliness" (note the different spelling) was a word Johnny Cash used in a song: "I'm giving you my onliness; come give me your tomorrow." I like the juxtaposition that onlyness is that which deeply connects you.

arranged marriage was now apparently a done deed. My future husband, a widower, was to give my mother a house, and she, an Indian
divorcée with three children, would no longer have to worry about
her finances. These tumbling sentences pronouncing my entire future
were both a relief to hear and the start of a new chapter of my life.

Even though I had been raised in America since I was nearly five, I
had always understood it was my duty to marry in this fashion. I don't
know when or how exactly that message was conveyed, but it was
clear that this would be my fate. I respected my mother enough to
want to do this for her, for our family—to do "right," especially given
all the sacrifices she had made as an immigrant to bring us to America
and the better life it promised.

Still, I did have my own private yearnings, so I turned to my progressive uncle, Zafar, for help. Zafar had come to America to attend
college and was now an executive at a big pharmaceutical company in
Palo Alto. He had been the family's male representative at the marriage negotiations, as my father had long been out of the picture. "Does
he [the groom] know I want to go to university?" I asked him. I was
attending community college, having deferred my entrance to the
University of California at Berkeley for a year. "No," my uncle replied.
"Your mother would not let me bring it up. You can discuss it with him
after you are married."

The implications of this went rattling around in my brain, as it
would mean a delay of at least a year, maybe more, before I could begin university. As is traditional in arranged marriages, I would not be
allowed to get to know the groom before we were wed. We might say
hello, or sit in the same room together in the company of others, but
he would not have any notion of my life goals, nor I of his. To tell
him what I wanted for myself would be quite out of the ordinary—
essentially impossible.

Given that was the case, I argued with, and may even have whined
to, my uncle. I knew from listening to the aunties' chatter that my

proposed groom had a housekeeper, a cook, a nanny for his child from his first marriage, and a big house on the hill nearby, so he was certainly not marrying me to help with the household. I felt confident that he would agree to my getting an education if it was understood up front that that was my intention. But if I had to wait and slowly build a relationship with him to the point where I could ask such a thing, the outcome would be uncertain. Certainly the one-year deferment to Berkeley I had gotten without my mother's permission would expire. Would he even let me keep going to community college?

Of the many things I value, education stands apart. In my own culture, it is too often reserved for the boys, who are expected to make the decisions. But I wanted that equal footing, and with it the ability to direct my future.

But the conversation was closed, my uncle stated. I could tell he was on my side, but he was also powerless in the dynamic. "There is nothing to be done," he pronounced.

My spirit was sputtering, and any gumption I had was dissipating. Over the next few hours the activities in my house wrapped up and everyone left. But I had one more—desperate—plan of action.

In a theatrical display, I told my mother that I was leaving home and began packing a disproportionate ratio of books to clothes into a cardboard box—but no toothbrush. It didn't matter; these things were all just props in this little drama. As she stood in my room watching, I exclaimed, "I am the product, and the deal can't happen without me!" *Just promise to ask him about my attending school*, I insisted as I departed.

Sometime that afternoon, after eating an apple fritter and a donut hole, my cardboard box in hand, I trudged over to De Anza, the community college campus where I was going to school, to use the phone.

By now, my mother couldn't construct full sentences, but *oh*, could she raise her voice. Ours was the dialogue of the deaf; neither could hear the other. So I hung up and called my seven-years-older, married sister. "Mom says she's going to kill herself if you don't come home," she warned. That's when I knew that this wasn't going to be easy. Even after that conversation with my sister, I figured I'd be home by the following morning, or maybe the one after. But that day never came, and nearly thirty years have passed since then.

THE CHOICES THAT DEFINE

In a matter of hours, my life had changed irreversibly. Everyone in my family called me disobedient and disrespectful. It hurt, but it was also true. But I was at least going to have the courage of my convictions. I had claimed what mattered, with both feet and a madly beating heart.

Looking back, this experience is when I first learned something crucial about life: Choices define us. The hand we're dealt is just a starting point; it's our choices afterward that reveal what genuinely matters to us.

How many of us have faced such situations—crossroads where we encounter immovable forces, where we must decide between making someone else's choice or our own? It's understandable how anyone can let the pressures of given situations, circumstances, or people around them define their next step.

But it's powerful to make your own choices, find your own path, open up your own door of opportunity—not just for yourself but for your purpose.

In my case, I was fighting not only for my own education but also implicitly for the value of education for everyone. I had plenty of examples in my life in which education was the key to opportunity. My uncle benefited from it, and even my mother was a positive example: As a divorcée in India, she didn't have the right or capacity to earn money and raise her three kids on her own, and so surely would have lost us. Leaving her children temporarily in different cities with different people, she moved alone to America, where she lived with my uncle and studied for a two-year degree in respiratory therapy.

My biggest concern at that particular moment, though, wasn't advocating for a big, purposeful cause but whether, in pursuing my "only," I would wind up being deeply lonely. My family, after all, had just ejected me. As it turns out, I found another family at De Anza: fellow teachers and students on campus who, like me, believed that an education is critical, whatever one's gender, age, religion, or heritage. The school's admissions officer, Lew Hamm, helped me find a few jobs with flexible schedules. I ushered at theater performances, did bookkeeping at the on-campus history museum, and coded a program in BASIC for the matriculation department. The support of those around me, far too many twenty-packages-for-a-dollar ramen meals, and the fact that community college courses were incredibly cheap kept me going, slowly but surely.

Discovering others on campus who, like me, valued education as the gateway to opportunity provided a sense of solidarity. Most were likewise piecing together jobs into a sufficient scaffold on which they could build more possibility into their lives. I took great comfort in realizing, finally, that I was no longer "the only one." I would later see this pattern repeated hundreds of times in business and in life: Finding "your people" sometimes means having to walk away from places you don't fit in rather than trying to squeeze yourself into a too-tight space with the aim to belong. It isn't *until* people make that choice that they are able to find others who share something meaningful with them.

It turned out that signaling my purposeful passion to the world led to the chance to make a dent when I was invited to advance the concept of educational access for all. At the age of nineteen, I was appointed by California's then governor, George Deukmejian Jr., to help "reform"—actually, radically reinvent—how California's 105 state community colleges worked. From 1987 to 1989, I served as the student board member on a committee that consisted of commissioners, judges, business executives, and state senators who sponsored the passage of reform bill AB 1725,* which transformed community colleges from

* Professor Tab Livingston's "History of California's AB 1725 and Its Major Provisions" can be found here: http://files.eric.ed.gov/fulltext/ED425764.pdf.

two-year vocational trade schools into an entryway to higher education. It made community college credits transferable, thus carving a
foothold for students to climb up to the next level.

Up to that point, attending community college had been a negative
in my mind. I thought it made me less than those high school friends
who had gone off to Harvard, MIT, or UC Berkeley. Yet this particular
experience turned out to be the key to my unique contribution. I
would later come to understand that it's not the "perfection" of your
experience that prepares you or earns you a seat at the table but rather
simply what you, and only you, can contribute—what you, and only
you, have seen, or what you, and only you, know. So many aspects of
my profile—my education level, my age, my income level—were unlike those of the other members of the college-reform committee, yet
it was precisely those elements of who I was that helped me contribute
something fresh.

So here was an early foundational lesson in how to make a difference: Bring what you distinctly have, align with others who share your
purpose, and make it happen, together. The problem is that this approach wasn't a doorway of opportunity that would always be propped
open, but a lucky, random coincidence that granted access that one,
singular time.

What most people do to give their ideas a solid shot is to climb high
enough in an organizational hierarchy to direct change. That's what I
did next—but not, as you'll see, particularly well.

SOMETIMES YOU FAIL SO YOU CAN REMEMBER

When Carol Bartz, who would later become CEO of Yahoo, described the
job arrangement I was being asked to take on at the software company
Autodesk in 1997, she called it "two peas in a pod," with me operating as
the "complementary half" of the Americas vice president, Michelle Pharr.
Michelle would be the market-facing executive, while I would serve as

the "revenues manager" of this more than two-hundred-million-dollar business. Convincing me took only Carol's assuring me, "You'll be among my top-one-hundred leadership team, you'll fix anything that can and will go wrong, and you can steer every new strategy to drive our market growth."

Even though my former boss at Apple, John Osborne, warned me that this arrangement would end badly, it seemed perfectly suited to me. After completing community college in 1989, I had advanced from an administrative assistant position to running a major-channel program at Apple. In 1996, when Steve Jobs famously returned to Apple, I left to join a start-up called GoLive (later bought by Adobe), where the business had grown to $4 million annually, as vice president of sales and channel marketing. When the Autodesk opportunity came along, I was between jobs, and though I felt competent at some things, I did not consider myself to be capable and proven yet in many things. This would be my first leadership role in a big organization, custom created by one of the few female CEOs of a Fortune 500 company. My ego heard this offer as "you are finally *enough*," and of course I jumped at the chance.

My mistake was in judging only what that role would mean for me and overlooking what everyone else would think about such an unconventional position. I didn't think about how a finance director might feel slighted by my having a title of "revenues manager." I didn't stop to wonder if the talented head of sales operations might be threatened by my oddly overlapping responsibilities. The results would speak for themselves, I reasoned.

As it turned out, my hiring quickly caused a kerfuffle. Human Resources was called in and a complicated "matrix model" written up to explain when the staff should come to me and when they should go to Michelle. Carol's promise that I could fix problems turned out to be true—any time people didn't agree, I was called in to mediate. I would learn later the many nicknames I had been given, including "the fixer" and "the queen's taster," but I approached my duties with passion, wanting to prove I was worthy—worthy of this job, of big ideas, and of the opportunity to sit at the table.

At one point, while preparing a multiyear growth strategy that would be presented to the board of directors, I disagreed with the marketing lead on how to spend the following year's budget. We disagreed privately, then publicly. I spent my energy making sure my argument was tight and then lobbied others to point out the flaws in my counterpart's argument. I did some serious spin.

Our disagreement eventually came up at the executive meeting. Instead of saying something neutral, I used what in the business world is the equivalent of the nuclear option: I warned that about $40 million of revenue was "at risk." These were effectively code words informing the CEO that she would fail because she'd miss Wall Street quarterly expectations. I then advocated for how the marketing budget *should* be allocated—in other words, my plan—thus throwing the marketing lead under the proverbial bus.

After the board meeting, Carol called me in for a chat, during which I fully expected to be congratulated. Carol agreed that I had done what it took to get results. But, she pointed out, I had alienated the team in the process with my negative personal machinations. There was no denying I was in the wrong, on so many levels, but I rationalized my decision by stressing the importance of the bottom line and arguing that having the *right* idea mattered above all. Carol countered that the way I had gone about winning meant not only that the team would never trust me again but that they wouldn't execute the plan because of how it came about. Regardless of how "right" the idea might be, it was effectively worthless.

I was fired soon afterward. It was my biggest professional failure.

It was also an epic personal failure. The person I had argued with, the person I had taken down, was a friend, whom I had worked with at Apple for many years before Autodesk. We were more than just former colleagues; we had trained for and run marathons together. When

I approached the finish line of the California International Marathon, the crowd was chanting my name, and I realized it was because of my friend: She had finished her own race some thirty minutes earlier and had gotten the crowd involved so I could finish strong. It was an act of generosity and love.

Of course the team would no longer trust me. I had been willing to take down a friend in the name of winning.

LOSSES AND LESSONS

In my departure from Autodesk I lost not only my job but also my conviction that being right, or having the right idea, was everything. In the deep hole of loss, I had only questions.

Clayton Christensen, the business guru, says that without new questions, there's no place for new answers. Where I had once believed that the theoretically best idea was the only goal, I now began to ask *how* to get the best results. Ten years later, after asking refined questions and finding clarifying answers to them, I felt I was able to reconcile the fundamental tensions inherent in collaborative work. There were many right answers, not just one. It wasn't enough for one person to advance an idea by a thousand steps—better to have a hundred people co-own an idea as their own so that each could move it forward a hundred steps. Co-ownership of an idea is what leads to successful execution, which, in turn, transforms that idea into what matters: a new reality. I wrote my first book, *The New How*, to explain what I had learned. I finally understood, through deep practice and shipping nearly a hundred products with many teams, that the future is not created but, rather, cocreated.

An idea is simply a steering wheel pointing to the future, and as such needs thrust and fuel to be realized. Thrust comes from the deep ties formed by many others wanting that idea to become reality, and their collective action is the fuel that moves the idea into the stratosphere. These facts would come to inform my understanding of how onlyness could work.

The final piece of the puzzle became clear and meaningful when my husband* asked me a rather strange question.

HONEY, CAN I QUIT MY JOB?

Most of us ache to make a difference—to leave the world better than it was the day we were born. But most of us also have jobs and commitments. So I was completely floored when my husband, Curt, announced one day that he wanted to quit his job as a technologist to get a PhD and focus entirely on what he could only describe as "world peace."

While his boss at the time was difficult, for the most part Curt seemed to genuinely enjoy his career. He explained that he wanted to do meaningful work, finding ways to help the people of developing nations help themselves. He was sheepish about his lack of clarity in his goals, mumbling about the inadequacies of the global economic model. Meanwhile, all I could think about at that moment was the personal economics of this decision. There were the not-so-small matters of our having one more kid to raise and put through college, our Silicon Valley mortgage, health insurance, and a myriad of other financial responsibilities.

If I was making a movie of that moment, it's here that I would insert a flashback to Curt at about ten years of age.

Because it wasn't my forty-seven-year-old, accomplished technology/chief architect/executive–type husband trying to tell me something. It was, rather, the kid my husband had been, who liked not only to solve puzzles other people created but to design complex puzzles himself. He was a kid whose family of eight siblings was the product of three different fathers. His school nickname was "Curt Dirt," because he was often neglected and because the household's dirty dishes often piled up by the back door instead of being cleaned in the kitchen sink. He would grow up and buy one sister a home to stop her from

---

* Yes, I did get married—twice, in fact, and both by choice, not by arrangement.

being homeless and help a brother to fix missing teeth to improve his job prospects. Curt had always noticed that people sometimes needed a scaffold to help them rise. He didn't seek to offer charity but to enable others to build solutions. Having "made it" despite many obstacles, he simply wanted to help others by contributing everything he knew about technology and the value of sharing information.

I wish I had seen all the personal meaning in his question at that moment. It's so obvious now. Before he embarked on such a venture, I suggested that he get a career coach, who asked him, "Why wait? A PhD will take years and won't really make much of a difference." Volunteering at a local food bank was probably what she had in mind, but as a systems thinker, Curt was aiming to make more of a global impact—a *real* dent with a measurable difference. He imagined doing something that would improve the lives of thousands—even millions—of people around the world.

He did follow his coach's advice by beginning with a simple act: blogging about what he cared about most deeply, namely, ways to improve conditions for people in economically challenged developing areas.[1] Even though he had virtually no readership, the blog enabled him to develop and express his own ideas more clearly. Next, he created a collaborative website (a "wiki")[2] where he could share his passion with a like-minded community. He had no idea if such a community even existed, but he dared to hope and acted accordingly. After a few weeks, Chris Watkins, an amateur importer/exporter in Indonesia, discovered Curt's writing and invited him to join an existing three-month-old wiki that was slightly further along in the same direction. Lonny Grafman, an instructor at Humboldt State University, had started that wiki.

No member of this motley crew was a leading expert in the field of development; what united them wasn't education, expertise, or credentials. Instead, they had something else important in common: a shared purpose about which they were equally passionate. Purpose can achieve something that money alone cannot, as it motivates the best *in* people and brings *out* the very best people.

ARMCHAIR ONLYNESS IN ACTION

Within a few months, Curt and his partners started to mobilize, form-ing a nonprofit, the Appropedia Foundation, and building the website Appropedia.org. The purpose of the foundation was to gather, orga-nize, and make available useful information about how to solve such real-world problems as obtaining clean water and reducing the spread of malaria. They had discovered that a large portion of projects (as much as 50 percent[3]) was actually repetitive work, looking for solutions and answers that already existed. Appropedia made the necessary know-how accessible. People could modify it for their needs, share their expe-riences, and find others with the same interest. In short, an extended network of people could now achieve real results faster and with less effort, without requiring anyone to be "in charge" to make it happen. The Appropedia founders still have day jobs and have never received a salary from their organization. Each of them spends several hours a week on the project—less time than most people spend watching TV. The funding for server fees and such is largely covered by personal do-nations and some small grants.

The Appropedia team's success in making a difference while devot-ing only a few hours a week struck me then as a crucial factor, as it meant that, enabled by modern technologies, anyone could begin mak-ing a dent in the world. At first, I thought of this as a way to shape the modern understanding of jobs and labor. But then I realized it could also change how any one of us could do something that mattered. Wit-nessing modest effort result in something significant and transforma-tive showed me the power of what is now possible.

In the nearly ten years of Appropedia's existence, it has enabled many people to create outcomes that matter: increased crop yields in Uganda, the use of solar power to improve water sanitation in rural India, the prevention of deforestation by the use of better, more effi-cient cookstoves in southern Sudan, and the construction of safer play structures in New York City. Today, Appropedia.org offers information about more than twenty thousand solutions, helping more than twenty million visitors a year and improving lives around the world. The thousands of people around the world who have participated by translating the site's content into different languages or adding their own solutions were doing so not as passive followers of some directive but as active agents who cared deeply about the issues involved. This type of following represented a new way to scale that I'd never encountered before. Previously I had only experienced scale in the corporate world, where scale was often about sameness: The ability to deliver hamburger number one thousand at the same quality level as hamburger number one was accomplished by routinizing tasks so it didn't matter who specifically performed them; people were simply cogs in a machine. Appropedia grew in size and impact not by asking any person to do the known thing over and over but by enabling each person to add their bits of what only they could, and have it add up to something meaningful.

For Appropedia, scale happened because people passionately took responsibility for whatever they could. Instead of seeing chaos, which I might have expected, I watched in awe as an entirely new system made things happen. *Seemingly powerless people, fueled by their deepest— even sometimes unnamable—sense of meaning find those who share a cause or purpose and act together, without needing to be told what to do, to make a dent.*

Without permission or needing to be appointed by someone.

Without specialized expertise.

Without a ton of money.

Without the external credentials of titles or education.

Without investing loads of time.

WHY NOW, REALLY?

Think about what's new and different in the Appropedia story: Could Curt and his colleagues have achieved what they did until recently? Likely not.

Each big technology shift changes the structure of society, and the Internet is just the latest. Individuals today have access to information that was previously available only to the establishment, if it was available at all. Now everyone can find, friend, or follow people who share their interests and passions using a variety of social media, whether Twitter, Facebook, or LinkedIn. Where once geography or social class limited us, we can now write to anyone and say, *I've been following your work and I have an idea for how we can create value together and get something done.* We can now raise money for a worthy cause on Kickstarter or Funding Circle. We can share ideas through near-ubiquitous Internet, mobile, and 3G access, via vehicles like Dropbox or wikis. Even language no longer presents a barrier, given the widespread availability of free translation tools.

Technology is the mechanism that now makes it *easy* enough, *fast* enough, *cheap* enough, and *efficient* enough to gather together people who share a purpose and galvanize them to act. It has also made possible a major shift in how we organize to create scale. Before the rise of the Internet, most people had to join a large organization—business, military, religious, governmental, or nonprofit—to attain the position and power that could enable them to change the world. That typically meant "fitting in" to a given organizational culture. New, fresh, and original ideas—born of different "onlys"—tend not to thrive in these contexts. Despite all the love talk in our culture toward disruptive ideas, they are more often viewed as unruly and untamed and so are easily targeted and killed within organizations, which also tend to avoid hiring genuine rebels, misfits, and black sheep.

Today, networks offer a new way to get things done. Any collection of people can pursue ideas together without organizational authority or hierarchies.* When value creation is institutional and hierarchical,

---

* How does one define "network"? Some people make it really complex. I prefer how Joel Podolny and Karen Page defined it, as any collection of people ($n > 2$) that pursues things together while lacking organizational authority. Published in 1998, their article "Network Forms of Organization" (*Annual Review of Sociology* 24:1–554) is a good resource for this model.

the vast majority of people are treated as cogs, dispensable and replaceable. But when value creation is networked, the distinct ideas, judgments, and decision making of individual players—you and I—matter more as the fundamental building blocks of value creation. This seems the ultimate game changer: While organizations and hierarchies continue to serve many useful purposes, we no longer *need* them to attain big goals.

MORE WHERE THAT CAME FROM

Watching my husband do the seemingly impossible from his red-checkered armchair made me ever more aware of other seemingly powerless people achieving comparably remarkable results. And the more I observed, the more clear the fundamental scope became.

Could it be that there was a new power at hand?

A new way in, even for seemingly wild ideas?

At last, a path for ideas born of our "one wild and precious life" to have a shot?

For many years and in the course of two books, I've explored how breakthrough ideas are discovered and valued, and why doing so is important. My first book examined how that process occurs within the context of firms. My second argued that value creation could increasingly come from outside the perimeter of the organization, so the key to success wasn't being competitive but, rather, being collaborative, including ideas from anywhere. That's where I first introduced the onlyness thesis, arguing it would be centrally important in an ideas-based, creative economy. So I have long been advocating that ideas do and can (and should) come from anywhere, and from anyone.

In *The Power of Onlyness*, I've pursued a new set of questions. First, instead of continuing to study firm-centric examples, where I was effectively pleading, cajoling, and convincing leaders to let new ideas in, I've now focused on the compelling forces of change: the ideas themselves. This led me to different domains, including education, filmmaking, coworking spaces, social justice, and, yes, even business. Despite dropping the jargon of innovation, the stories that follow embody innovation to their core by showing exactly how new ideas turn into new realities. Second, because the power of the person bringing an idea can either *liberate* or *limit* that idea, how to change those power dynamics became key to understand. That meant asking how new ideas can emerge and scale despite the relatively low power status of those who have them. What are their strategies, and what are the precise reasons why those strategies work? And how can others emulate them?

What became clear after researching nearly three hundred dent makers is that there *is* a new path forward. Studying these people taught me a lot about how ideas can start out small yet make a huge dent. This book tells twenty or so of their stories. These accounts are personally inspiring and, at the same time, highly instructive. Each illustrates useful principles and practices, independent of story and place, and my hope is that together they will serve as a guidebook for you that is both practical and motivating.

All of these stories reveal how ordinary people who would once have been unable to make a dent in the world have done so by acting from their purpose, finding meaningful allies, and then mobilizing many to act as one. Among them are:

- André Delbecq, a former business school dean. Delbecq pursued a question that was downright heresy in the business curriculum: Can faith be taught alongside the pursuit of fortune? His quest instantiated a completely new field of study: spirituality in business.
- Franklin Leonard, a twentysomething binge-watching film lover who began a Hollywood career at the lowest level, yet got the entire industry to reconsider how scripts were picked to become films.

- Talia Milgrom-Elcott, who with a staff of only ten has managed to mobilize hundreds of organizations to train a hundred thousand STEM (science, technology, engineering, and math) teachers in ten years, establishing the kind of learning and risk-taking context required for people to solve previously unsolvable problems.

While each story is different—spanning young and old, low tech and high, novices and experts, ranging across the world and a variety of subjects and types of dent—each demonstrates how, because of onlyness, new ideas end up getting their shot to make a difference. No matter what the person's age, or gender, or color, or so on, the ideas had a shot. To reshape industries. To advance agendas. To right wrongs. To invent things. To address age-old problems. To simply get things done. These stories not only give us hope but show us a pathway to the future.

Wild dreams now end up making a difference. They show us that it is possible for each of us to do the same, but only if we know *how*. This is where *The Power of Onlyness*, this book, fits in.

PART I

Your Dent

The Power of Your Own Meaning

This book is organized in three sections.

- Part I, "Your Dent," explains why your "only" matters and how it is not a path to loneliness but instead the way to be powerfully connected.
- Part II describes the power of *us*, not only how to find your co-denters but why having this tribe to belong to is crucially important, for your ideas to have a shot.
- Part III, "Denting," explores how many can act as one to achieve productive results without sacrificing the individual meaning that starts the journey.

While onlyness is not about your personal brand, or even your ego, it does start with you. To build and use networked power, you must organize around a shared purpose, a common meaning. So until you understand what matters to you—and, more importantly, *why*—you can never really and truly command this power.

In chapter 2, the example of Kimberly Bryant reveals why you must own *all* of you, even the parts others disparage, and what it takes to make meaning from your history and experiences, visions, and hopes. I'll also share a personal experience that shows why the brand of you isn't the point. The account of Zach Wahls and his friends illustrates why onlyness isn't lonely but uniting. In chapter 3, the story of André Delbecq demonstrates how to discover your own power by chasing a question even if you don't know where it leads. Charlie Guo shows us why focusing on oneself, and not one's purpose, limits what is possible. And Cindy Gallop shows how the simple act of speaking one's own truth is a powerful way to find your dent to make.

All these stories show how real people in their everyday lives go about adding their value to the world. They aren't polished-up, glossy magazine–type accounts but stories of people like you and me: complicated yet courageous, flawed yet perfect, in progress and yet already making a dent. Because they've groped their way along, fumbling in the unknown and dark spaces, falling over things as they progress, experiencing bruises and bumbles, their stories offer us a light switch and ways to navigate our own spaces.

CHAPTER 2

Starts with You

Fitting in is the greatest barrier to belonging.
—BRENÉ BROWN

KIMBERLY BRYANT: REDESIGNING THE GAME

As she dropped off her twelve-year-old daughter at a coding camp for gamers, Kim Bryant became aware that, despite the bucolic setting of the Stanford University campus, things might not go smoothly for Kai. Although there were thirty kids in Kai's class and two hundred in the overall program, Kim was able to count the total number of girls enrolled "with my fingers and toes." And, she noticed, "the class was pretty white, too. Lily white."

Her response to this observation started Kim on her dent: to change the world of tech to include more people like her daughter.

"Follow your passion" is the advice that is so often given to new graduates, job seekers, and aspiring entrepreneurs. But what if what you care about is something that seems to concern no one else, or seems too small or inconsequential? Do you ignore it? As we'll see in

Kim's story, to make a big dent, you don't have to have a big idea. Your idea can be quite small, or even dismissed by others—but that doesn't mean it doesn't matter.

THE ONLY ONE

Talking over dinner at the end of the weeklong camp, Kai told her mom that the teacher gave a lot more attention to the boys and that many times her own questions were dismissed out of hand. "They didn't mentor or encourage the kids in the class equally," Kai said. Already a passionate gamer, she had lots of ideas but was treated as an irrelevant novice.

"I'd seen a light in her eyes that I didn't want to see put out prematurely," Kim recalls. Like all parents, she wanted to encourage her child to follow her interests, and show her that her questions mattered.

That night, after Kai was in bed, Kim's engineering mind got busy thinking about ways to address the situation. Her first thought was to send a few of Kai's friends with her to change the ratio at the next camp. It wouldn't be cheap, but Kim's tech job enabled her to afford the fifteen-hundred-dollar-per-kid tuition to keep Kai from being the "only one" in her situation. Then, Kim had another, even better idea.

Before we get to that, though, ask yourself this question: What would you do if you were faced with Kim's situation?

You could simply ignore it. You could conclude that engineering or coding might not actually be a good fit for your child, and if it is, she'll find another way in. You could seek an alternative, either in a similar camp elsewhere or a different type of camp altogether. You could volunteer at the camp and attempt to change its approach and equally encourage *all* kids who enroll, regardless of gender or race.

How you respond to a situation like this—a challenge, a new situation, something that bothers you—depends very much on who you are, what you care about, and why.

So let's understand who Kim is, through a few stories.

I first met Kim at a conference on a cold day in Boston, and she was talking about southern cooking. She described herself as a "down-home girl," and when I asked her what exactly that meant to her, she responded with a question of her own.

"Do you know Memphis?" she asked. "Growing up there had an indelible impact on my life because of its history, a history that is not so pretty. MLK's assassination in Memphis made me realize that there are some things worth dying for. And, to be clear, I'm from North Memphis. Not many role models there. Not too many kids got out."

Our origins are part of our heritage. This is not to say that we're defined by our pasts but, rather, that the context of our particular background shapes what we notice and respond to, which in turn makes us who we are today. Kim's "only" has been deeply influenced by growing up in North Memphis and what she believes is worth fighting for.

Kim beat the odds by being accepted at Vanderbilt University as an engineering major, which ultimately led her to experience being portrayed as a rebel on the front page of a national newspaper.

"The thing is," Kim explains, "I was almost always the only brown or black person when I went to school. Almost always the only woman in class." She loved math, science, and related subjects yet felt deeply alone in those courses, which may have accounted for the fact that her freshman-year grades were dismal—culminating in a 1.3 grade point average (GPA).

But she found friends in the black student union, which was "one of the few places on campus where I felt I really belonged." As a result, Kim got involved in student activism. Vanderbilt was still invested in South Africa, and she and her friends wanted to end apartheid, so they took part in protest marches, spoke to provosts, wrote a manifesto, and implored the president of the university to make changes in the school's policy.

At one point, USA Today and the Wall Street Journal showed up on campus and took photos, and Kim's activism was displayed on the front page for her mother to see. "My mom was seriously worried," Kim notes. "She kept thinking my scholarship might get revoked." Throughout Kim's life she had been advised not to challenge convention because someone might be upset by it. Her grandmother straightened her kinky hair at the stove with a hot comb so she could look more "appropriate," more respectable. Doing only that which others expect of you, or from you, was a strategy taught to Kim to protect her. She was torn between her desire to be a part of the solution to the injustice she saw and her mother's advice to conform. "I heard her, I swear I did. But I also knew it was important to say something, to do something." Her efforts paid off, as Kim and her friends ultimately swayed Vanderbilt's administration. It was one of the first schools in the world to shift its investment strategies in response to apartheid, joining with others deeply connected in purpose to foster change in South Africa.

Kim's success gave her resolve on the academic front. She hit the books, making up for lost time. By the time she completed her BS in electrical engineering, with a computer science minor, she had earned enough As and Bs to boost her GPA from nearly failing to a solid B.

TWOFER

After Kim graduated, she landed a solid job at Westinghouse, and then an even better one at DuPont. She remembers thinking she had "survived school, mastered college, and then *arrived*, in a capital-A kind of way, in a profession my mother could only have dreamt of for me, and that I would *finally* have a peer group and would feel less lonely." She was the exceptional person who "got out" of North Memphis; she thought it meant real change.

Her DuPont manager introduced her to the team by saying: "With Kimberly coming on, we got a twofer," meaning that they hired both a woman *and* a person of color. Her deepest hope that her ideas, and not her color, would matter was already being squashed. What had been highlighted was her being different in a way that simply limited what was possible for her.

The effect of being the "only one" is well known. In 1977, change expert and Harvard Business School professor Rosabeth Moss Kanter did a study that showed that when individuals are members of a group that represents less than 15 percent of the total, they experience three things that constrain their ideas.* First, they feel highly watched and thus have a burden of performance pressure. Second, they feel isolated and excluded from social settings (which is where relationships and trust are built) that would enable them to succeed. Third, they feel tremendous pressure to assimilate to the group's norms. Kanter described this tokenism as difference being allowed in, but in such a small amount as to never affect the power structures.[1] People who are "firsts," "onlys," or "the exception to the rule" feel this effect. And what it means is that they are more likely to adapt to the context rather than dent it to reflect their own ideas.

While demographics like age, gender, race, and sexual orientation don't negatively affect the quality of ideas, being a token in an organization does, because what is noticed first is *different* rather than *that idea that we need to study to see if it can take us into the future*. It's no surprise, then, that Kim was so sensitive to Kai's being the "only black girl" in the coding camp. Nearly twenty years have passed since the "twofer" comment, and little has changed regarding women in technology. If anything, the situation has only gotten worse.[†]

WHEN DO YOU SCREAM NO?

Although Kim's introduction at DuPont underscored her separateness, she was still very proud to have been hired there. Her employer provided some good benefits, including mentoring for new hires. Kim's

---

* Rosabeth Moss Kanter, "Some Effects of Proportions on Group Life: Skewed Sex Ratios and Responses to Token Women," *American Journal of Sociology* 82, no. 5 (March 1977): 965–90. The article is publicly available and free to read at http://www.jstor.org/stable /2777808?seq=1#page_scan_tab_contents.

† For the most part, the percentage of computing occupations held by women has been declining since 1991, when it reached a high of 36 percent; it's in the midtwenties now. Details are in this report: Catherine Ashcraft, Brad McLain, and Elizabeth Eger, *Women in Tech: The Facts* (Boulder, CO: National Center for Women and Information Technology, 2016), https://www.ncwit.org/sites/default/files/resources/womenintech_facts_fullreport _05132016.pdf.

assigned mentor was another female engineer about ten years older than her.

After Kim had been at the company for a while, she identified a role she wanted to move into: "It was only a rung or two above my relatively entry-level one, but I could see myself growing in that direction." At her next mentor meeting, she brought up the idea. "I want to be an area manager in a few years. What do you think I should be doing now to prepare?" she asked. It was a standard, well-reasoned question to ask of a mentor. She hoped she could get advice on what to study, what soft skills to develop, or how she might volunteer in her spare time.

Instead, her mentor told her, "You can never achieve that. No, not *you*. You will not be that."

Surprisingly, or perhaps entirely unsurprisingly, this only made Kim more determined. But it wasn't just her resolve that mattered. Her ability to choose a response—to decide for herself what to accept and what to deny—was a key skill, not only for navigating her life but also as a way to unlock her capacity. To choose a response is to reject the belief that we are at the mercy of others and not in control of what happens to us. While we may not be responsible for much of what occurs in our lives, we are completely responsible for our *reaction* to it.

"It only made me want to prove her wrong," Kim recalls. "It was crushing at some level, but at the same time not crushing because I refused to believe it as true." She vowed to herself, "I will be bigger than *you* can imagine for me; I will be as big as I can prove to myself to be." Within five years she had actually risen beyond the level her mentor had told her was unachievable. She clearly remembers reaching that higher position in the organization and thinking, *So there.*

Often, when our inner voice loudly screams *no*, we learn something about ourselves. We glimpse a vision of what else is possible, even if it's something that's never been done before.

KIM'S DENT

These four aspects of Kim's character—Down-Home Girl, Rebel on the Front Page, Twofer, and Saying No to the Mentor—inform and shape how she came to her dent: how, in 2011, she enabled Kai to learn coding when her daughter didn't fit in to the cultural pattern or bias surrounding what coders look like.

She began by designing course work in her spare time, then borrowed old and used laptops so she could bring Kai and four of her friends together to encourage their curiosity and create a context for learning. With those first five kids, ages six to thirteen, gathered around the kitchen table, she began to close the digital divide for girls of color. Kai was not only supportive but was visibly proud of her mom. This project brought them closer together.

"At some level, I thought the situation Kai faced would be different than what I faced," Kim explained. "But there's still a dearth of African American women in science, technology, engineering, and math professions . . . an absence that cannot be explained by, say, lack of interest in these fields." In her design, Kim knew that it was not "enough to create change for just one person," because that would only perpetuate the existing culture. "The pipeline fails because of the leaking of that pipe and the unwelcoming culture that pipe flows into." Kim's own experience taught her that she needed to create change on a larger scale. "By enabling many kids, there are enough people like you joining alongside you, thus changing the culture."

It mattered to her as a mother, too, as jobs in STEM fields pay well enough to raise a family. As a single mom herself, and the child of a single mom, Kim is one of the 56 percent* of black women who are

---

* Fifty-six percent of African American women are single parents, according to the US Census Bureau, in 2015, as compared to 19 percent of white women. See "Table A3: Parents with Coresident Children Under 18, by Living Arrangement, Sex, and Selected Characteristics" at http://www.census.gov/hhes/families/data/cps2015A.html.

single parents. "Black women are the only things holding the black community together. Having a job that allows you to support your family on one income is the way you hold your head high. It gives dignity while also giving opportunity."

Even the name Kim chose for her organization was a sign of her intention to change the game. "Using the word 'black' scared me a little at first. Too many people see that word and see it as a negative," Kim says. She asked a fellow attendee at the BlogHer conference for advice. Analisa, a Filipina entrepreneur, appreciated the challenge immediately. Kim recalls the moment when Analisa "stopped shuffling what she had been looking at, looked straight at me, and said, 'It's what you're trying to do for your community, so *call it that.*'"

Don't let others define *it* for you. You define *it*—even if you need to *re*define it.

"Am I worried it's too bold?" says Kim about the name of the organization. She pauses a minute and then responds, "It needs to be. We sow new seeds, create new paths, and from now on, the rules are different. It's a new game; it's what we do, not what we hope to do."

The very design of Black Girls Code (BGC) opens up opportunities to celebrate the onlyness of all its participants. While it started out as a side project, as so many "dents" do, it quickly grew as other moms started chapters for their own girls. As of January 2016, the group has chapters in seven US cities as well as one in Johannesburg, South Africa. More than seven thousand girls from ages six to seventeen have learned skills like robotics, video game design, app development, and computer programming.

An estimated 1.4 million new coding jobs will be available by 2020,[2] creating a pivotal moment in economic history. "These are the most lucrative careers of our economies," Kim observes, "for which women of color and women in general are being bypassed. That has got to change." By 2040, Kim aims to teach one million girls to code.

Nyatche Martha of Orange, New Jersey, was twelve when she attended a BGC hackathon camp. The experience taught her that "you can put anything that is in your head, something that you see needs to be created, and then make that something that anyone can use."[3] Nyatche describes succinctly what happens when onlyness thrives: the very capacity to participate fully affects the outcome. New ways to address old problems are found, entirely novel ways of approaching issues change the status quo, and solutions to problems that were previously ignored can now be reconsidered and possibly make a dent.

One BGC participant designed an app to show incidents of violence in her city. Everyone who uses it—moms, teachers, other kids—can drop a pin on a map to show where groups of troublemakers are hanging out or where drug deals are going down, which helps all the other kids walk home safely. Another app created at BGC was an inventory system for local food banks.

Imagine a world where a lot more of us can solve the problems we see or build something only we can imagine.

Progress for one person ought to drive progress for many, but it so often ends up meaning that just one person succeeds, often despite the odds—like Kim, a "lucky one" who was exceptional. Having one person at a time triumph doesn't change things for the rest of us; it doesn't shift the status quo. We need a new pathway—indeed, a new *power*—for more ideas to count, not to be the rare exception. If only a few ideas are valued, the wealth of many humans is lost. The question of what is created depends on who has a place at the table.

Kim describes this phenomenon as "code or be coded": Either shape the world to include your ideas or be relatively invisible in it. Kim's unlocking what mattered to her led her to find people with whom she could make her dent, and that gathering of people in turn started to unlock the power of many others. This is the flywheel of change set in motion, endowing the world with more variety and creativity and inviting more of us to act out of our purpose to create a cascade of remarkable consequences.

CLAIMING IT

What can Kim's story teach us about how we find our own path?

First, claiming your spot is not just about recounting one's history and experience, but about defining what that history and experience have come to mean to you. What did you say no (or yes) to, whom did you decide to support or reject, and what did you decide to persist with, despite pressures to stop? When Kim rejected her mother's advice and kept protesting against apartheid, that was a decision. When she studied hard enough to start getting As in her engineering classes, that, too, was a decision. When she saw that Kai was struggling, deciding to do something to help her was a decision. If she had made the decision not to act, that, too, would have been a choice. These decisions define and form your puzzle piece into its own distinct shape.

Second, what surrounds you affects you. Your particular context teaches you about the world and shapes you, just as Vanderbilt, Memphis, DuPont, and the groups of people she met in those places set the context for Kim's only. It's popular, especially in Western cultures, for people to think of themselves as "self-made," but their full selves are actually built by and shaped by interactions with others.* Compared to any other being, we humans spend the longest amount of time in childhood,† during which we learn *from* others and how to become *like* others, benefiting from the inventions and creativity of previous generations. But those who share an identical context don't end up exactly the same. Nicholas Christakis, a Yale professor, has offered an analogy that explains how people can have similar experiences yet be so different.[4] He points out that both pencil lead (graphite) and dia-

---

* There is a lot of research on the topic of identity and social context. I'm drawing on Tom Postmes, S. Alexander Haslam, and Roderick I. Swaab's "Social Influence in Small Groups: An Interactive Model of Social Identity Formation," *European Review of Social Psychology* 16 (2005): 1–42, doi:10.1080/10463280440000062.

† This is a fact cited in Yuval Noah Harari's *Sapiens: A Brief History of Humankind* (New York: Harper, 2015), a book that is well worth reading.

monds are composed of carbon. But their constituent atoms have been exposed to different degrees of heat or pressure, which affects how they connect to their neighboring atoms. Similarly, another individual who might have had a similar context in Memphis or at Vanderbilt wouldn't have noticed what Kim did or interpreted it the way she did.

Which brings us to the third lesson of claiming one's only: valuing your full self, even those parts that others might disparage. You will need the confidence to say no to those who attempt to define something that is inherently yours. When someone tries to define your worth based on a group stereotype—in Kim's case, for example, her racial and gender identities—you must recognize it as a trap and resist. All of us confront stereotypical group narratives that have nothing to do with our actual individual passions and interests. A woman who is told she is "more collaborative" than men might find herself fighting her own nature if she's not naturally collaborative. A man who is expected to be the family breadwinner might repress how much he wants to be a stay-at-home dad. A young person who is told she can't possibly know enough to contribute might find herself frustrated because she in fact has a great passion and creativity in a given area. It takes a certain strength of character to value yourself for who you are, *as* you are. Growing up, for example, I was told my role—as a girl, an Indian, and a Muslim—was to serve my family, so when I wanted something for myself that my family challenged, it was nearly impossible to claim what I thought mattered.

And to those of you who have had things happen to you that you wish hadn't—your parents divorced, you were bullied as a kid, you've had cancer, your family was poor, your spouse abused you—you may feel a sense of unworthiness, that you (and therefore your ideas) don't matter. But these circumstances are not inadequacies. There may even be a beauty in the brokenness. Japanese potters use a technique called *kintsukuroi* ("golden mend"), a method of repair that involves a lacquer resin laced with gold, so that the site of the break is not hidden but made beautiful, the flaw becoming part of the object's worth. Through *kintsukuroi* the brokenness becomes a record of an event in the life of an object rather than the cause of its destruction.[5] Acceptance of what has happened is not the same as acquiescence; things can still be fixed. But to deny your own experiences, whether something happened to you or you made a mistake, is to deny a part of yourself, of your fullness, and thus limits the source of your creativity, ideas, and perspectives.

To claim yourself as whole is to assert your own value—not because everything about you is perfect but because it is all perfectly yours. This acceptance of your full self is nonnegotiable in claiming the power of onlyness. If you can't value what you alone have to offer, no one else can either.

Now, what do these three elements—choices, context, and claiming—have in common?

If we look again at Kim's story, we can see how she is shaped by context yet distinctly her own person, how her choices shaped her life, and how she claimed even those things about herself that others dismissed as negative, defining for herself what they meant.

Story	Meaning
Born in Memphis	Certain things are worth fighting for.
Not for You	You don't define me. I do, by doing the work.
Being Born Black	I am not defined by stereotypes, but by my individual choices/actions.
Get Out	Don't just escape the past, change the future.
Rebel on the Front Page	Together, we can create social change.

You see that Kim has defined a meaning for her history and experiences, visions, and hopes. The root of this power is not what happened but the meaning one defines from it.

I wrote down Kim's stories and then drew a through line; so can you for your own stories. It's worth noting that this narrative story need not be a "big" purpose or even the "final" purpose. The size or weight of onlyness isn't the point. Notice that it need not have a name, because that's not the relevant part—only that you can understand enough of what matters to you to understand how to then act on it.

When your life has meaning, it's because you have defined that meaning.*

John W. Gardner, the former secretary of health, education, and welfare under President Lyndon Johnson, said, "Meaning is not something you stumble across, like the answer to a riddle or a prize in a treasure hunt. Meaning is something you build into your life. You build it out of your past, out of your affections and loyalties, out of the experience of humankind as it is passed on to you, out of your own talent and understanding, out of the things you believe in, out of the things and people you love, out of the values for which you are willing to sacrifice something. The ingredients are there. [But], you are the only one who can put them together into that unique pattern that will be your life."[6]

FROM THAT SPOT WHERE ONLY YOU STAND . . .

With the focus on preparing the next generation of workers, a key topic of concern for our economy, and with feminism experiencing a new wave of interest, you might conclude that Kim's work has come at an ideal time in the culture. But it's worth remembering that she started this work about seven years ago, when almost no one thought it was an especially viable idea. Since then, *Marie Claire*, Oprah Winfrey,

---

* Sometimes people look for the thing that makes them happy rather than the thing that gives them meaning. While these can overlap, they aren't one and the same. Jennifer Aaker, Stanford professor and author of *The Dragonfly Effect*, studies happiness and meaningfulness. I turned to her insights on how they differ. She and her colleagues found that meaningful choices are often not pleasurable to make and indeed may come at a cost or involve pain; they are often associated with a larger purpose. If you want to study more about the distinctions, her research is well summarized in two publications: see Roy F. Baumeister, "The Meanings of Life," *Aeon*, September 16, 2013, https://aeon.co/essays/what-is-better-a-happy-life-or-a-meaningful-one, and Emily Esfahani Smith, "There's More to Life Than Being Happy," *Atlantic*, January 9, 2013, https://www.theatlantic.com/health/archive/2013/01/theres-more-to-life-than-being-happy/266805. Her other big insight? Happiness is more about getting; meaningfulness is more about giving. This explains why I say that onlyness is that which you have that can serve the world.

and even President Obama's White House have "discovered" her, and the reason most of us think her timing is perfect is because Kim and others like her had the foresight to dedicate themselves to an idea, and by doing the work, they shaped the conversation that followed.

Her experience offers an important insight on the origins of big ideas.

When Kim started Black Girls Code, she held a full-time job and did BGC "on the side." In fact, of the nearly three hundred individuals I studied for this book, most of them started the process of making a dent as a side project. Undertaking new ideas involves risking yourself. This is especially true if your own status (or lack of it) will make your ideas seem too *much* or too *wild*. Thus, side projects and extracurricular activities enable you to test out ideas safely. By exploring interests and passions without too much at stake, you can do the work without compromising current commitments, in a playful, low-stakes way. "On the side" lets you build something one piece at a time, much like putting up girders and spans for a new wing of a house that you want to later occupy.*

An obvious question to ask here is how one finds the time for side projects. Kim fits her work on Black Girls Code in on weekends, which means watching less TV, cutting back on surfing the Web, keeping shopping to a minimum, asking another parent to take care of her daughter for a playdate, and so on. When my husband began Appropedia, he woke up an hour earlier every day to attend to it. You can always find time for things that matter by consuming less and creating more.

The other major concern is where the money for a new endeavor will come from. Ask yourself how you allocate financial resources for things you need in your personal life. You first decide what it is you require, then find out what it will cost, and then figure out how to pay for it. Every start-up entrepreneur with a vague notion of an idea and a business plan typically asks for a huge amount of financing. More often than not, though, he or she doesn't have a clear idea of such

---

* The term "girders and spans activities," developed by organizational behavior professor Herminia Ibarra, was found in her first book, *Working Identity: Unconventional Strategies for Reinventing Your Career* (Cambridge, MA: Harvard Business Review Press, 2004).

basics as what equipment they need, what specific pilot project would prove out the thesis, or what kind of partners should be involved. Thinking big is great, but you must also undertake the hard and specific work of a proof-of-concept. Develop a concrete short-term plan and you'll have a much clearer sense not only of what you need from others but of what fundamental work needs to happen in order to make the long-term picture even possible. Kim borrowed equipment and pieced together prototypes to prove out her concept, and financing for BGC began to flow in *after* she had taken those preliminary steps and worked through the risks. Importantly, she didn't wait to understand how she'd make money, or even what the final outcome would be, as a precondition of undertaking the project.

Despite not having the time, or a big brand presence, or money, Kim brought her nascent idea into a new reality. How?

By standing in that spot *only* she stood. Even grand ideas may look ridiculously small at their outset because only a few people (maybe even just one) can see their value. Your act of noticing the idea that matters is the first big step on the path forward.

Begin by Noticing What Only You See

The idea expands as you find your co-denters and collectively grow it large enough to be observed by many. Convention says there's some magical "tipping point" at which this occurs, when in actuality dents develop in a series of moves, where choices lead to actions, which leads to readiness, which in turn leads to more actions, and so on.

"Give me a firm spot on which to stand and a lever, and I shall move the earth," Archimedes said. In onlyness: firm is arrived at by what gives you meaning, your spot is being entirely yourself, and your idea is the lever that moves the world.

PURPOSE OR PACKAGING?

What distinguishes what has just been described—expressing ideas that matter to you as a means of beginning to make a dent—from having a great personal brand? The concept of creating a personal brand has recently become a very popular topic. Blog this, Instagram that, stand out from the crowd, make yourself seen, get your ideas heard—every "expert" in the area advocates taking many of the same steps to define oneself.

In 1997, Tom Peters, a leadership thinker, published a magazine article titled "The Brand Called You."[7] I remember exactly what bench I was sitting on as I read the article, which put the world on notice that each of us was responsible for owning who we were on the deepest level as contributors in the new marketplace. Peters urged readers to "cast aside all the usual descriptors that employees and workers depend on to locate themselves in the company structure." Forget your job title, he said, and instead ask yourself: "What do I do that adds remarkable, measurable, distinguished, distinctive value?" Forget your job description and ask yourself: "What do I do that I am most proud of?"

Peters's work was a loud wake-up call for people to break free of the need to conform *in* and *to* their workplaces. He imagined a future much like we're experiencing today, where ideas and not organizations would be the central locus of value creation. Others would later describe this new landscape as the knowledge economy, the ideas economy, or the experience economy. This breakthrough concept changed the dialogue and perspectives around work and opened up a whole new way of thinking. It pointed us to what mattered: ideas born of personal passion that come from only *you*.

It's unfortunate that this insight, though, was expressed in the language of marketing, brand, and branding. Doing so conflated the wrapping with the gift it held. It's not that the wrapping and the gift can't be considered as a single package, of course. Branding is, at its best, a reflection of who you are.* And your reputation matters, because having others know what you care about enables you to build a trusted network who can galvanize action alongside you.

But to make packaging the priority is to distort Peters's key message: to discover how you can most effectively add value by stepping outside of limiting frames and doing meaningful work that matters.† This distortion happens all the time. I was reminded of this, personally and painfully, only recently.

BE EDGIER, BUT NOT TOO EDGY

For several years, at professional conferences, I would often run into a well-known business guru, who would take me aside, put an arm around me, and kindly offer to lend a hand with my career in any way he could.

As my second book was coming together, I thought he might be helpful in coming up with the title. At a seminar, I reminded him what I was working on, and he nodded as he recalled some of our earlier e-mails, in which I'd shared the key ideas.

Full of anticipation, I waited to hear his advice. "As a brown woman," he started off, "your chances of being seen and heard in the world are next to nothing. For your ideas to be seen, they need to be edgier."

---

* While it's entirely possible that something can be good *and* have the appearance of being good, it's also possible for something to be not so good, yet have all the appearance of goodness. Yeah. That's one problem with conflating the wrapping and the gift: Brand can be simply a better representation than what's inside the packaging. And yes, the "appearance of goodness" is a reference to Eliza Bennet's line "One has got all the goodness, and the other all the appearance of it," speaking of Fitzwilliam Darcy and George Wickham, two other characters in Jane Austen's *Pride and Prejudice*.

† Jennifer Aaker reminded me that it was her father, David Aaker, who helped develop the thesis that brand matters. While he was mostly considering corporate brands, he wrote, "When you develop [a] brand, you don't look at what is unique about the brand—you look at what you want to stand for, your value, and [that] starts with what your vision is. You don't really care if it is the same as others or different." Google his definition of brand equity and you'll find it's described as "the value you provide." So the early notion of brand was clearly pointing to what I term here the "gift." But it has become overused in people's eagerness to make sure they are known.

Then he paused to ruminate a bit, frowning, as he stared up at the windows of the former church building we were in. Looking up at the angels and archangels, he then added, "But if you are edgy, you will be too scary to be heard."

In short, I needed to be edgier to stand out from the crowd, but if I chose that option, then I would be too threatening to be considered as a viable business thought leader. Because of my "packaging," and not the merit of my ideas, I would have absolutely no chance. Brand, according to the advice he had given me, was just another construct that demonstrated how I didn't fit in.

At first, to be entirely honest, I was simply confused. I headed off to meet with some of the other seminar participants for dinner, hoping they could help me digest what I'd been told. But when I was met with blank stares, averted gazes, and several sighs, I started to become a puddle of anxiety. One awful, sleepless night later, my worry turned to fear. The question I kept asking was, *Is he doing me a service by telling me this?*

It took me a while to understand what had actually happened in our exchange. It's important to remember that this person wasn't trying to be mean; he was trying to be helpful. He was simply explaining the rules of power *as they exist today*,* the social constructs that shape and constrain who gets to bring ideas to the table. The limiting factor here wasn't a person saying, "You can't be successful without conforming," but rather the unquestioned and unchallenged frameworks of power, which were being accepted and then passed along as the "truth."

Of course, it *also* meant that the guru hadn't been focused on the

---

* To understand the existing frameworks of power, I turn to organizational management expert Jeffrey Pfeffer, who is a Stanford professor and the author of the definitive book on power, rightly titled *Power: Why Some People Have It and Others Don't* (New York: Harper-Business, 2010). In 2013, Jeff published an academic piece arguing that despite many new theories on power, the existing theories are still valid. That piece, "You're Still the Same: Why Theories of Power Hold over Time and Across Contexts," *Academy of Management Perspectives* 24, no. 4 (November 2013): 269–80, can be found here: https://www.gsb.stanford.edu/faculty-research/publications/youre-still-same-why-theories-power-hold-over-time-across-contexts.

new ideas, or the meaning of the work, or the people best served by it. He did not consider its potential value and the dent I was aiming to make. What he was focused on instead was how I would be perceived, my image, whom I would be compared to . . . all a packaging exercise. This is what happens when you conflate the gift and the packaging. It's so subtle and so commonly done that we may not even notice it has happened, and we might even think the person who points this out is saying that "marketing is yucky." But that's not it.

The reason to consciously be aware of the distinction between the two is simply this: It keeps your focus on what you need to be paying attention to.

Of course, there's the immediate issue that whenever you don't fit in to an existing category (or the category doesn't exist), your ideas and freshness can all too easily be lost in a packaging-focused world. In my case, the category of management thinkers has primarily been shaped by Ivy League–educated, white, Western, heterosexual older men, so people who look like me could understandably be encouraged to be something other than themselves as a way to "stand out." By keeping the focus on the gift itself, though, you can keep your ideas always at the forefront—even if they don't fall comfortably into an existing "category," even if there are many other people doing similar things.

Onlyness, your signature concoction of what matters to you, gives you clarity of purpose and enables you to focus on what matters. That will be our path forward. Existing power and status frameworks act as self-reinforcing loops, keeping the status quo the status quo.[8] Until these loops are broken, nothing changes. Not for you, not for anyone. Don't buy into the trap. You don't have to be self-consciously "edgy" or stand out; you simply have to be yourself and keep advocating for the ideas that matter, thus creating the value you aim to bring to the world.

And it's probably worth noticing that "onlyness," the word, does not represent *more* of you; it is not, as some might assume, a suffix of "ness" added to the base word "only" to make a new word that puts the emphasis on the idea of you.* Instead, it is a word that braids together how people create value in this modern economy: First, you stand in that spot in the world that only you stand in, then you meaningfully connect with others so that you (and, more to the point, your *ideas*) finally have a new pathway in.

---

* In his book *The Art of Language Invention*, David J. Peterson—the inventor of the Dothraki language for HBO's *Game of Thrones*—reminds us of the difference between a suffix and a circumflex. A suffix is a term added to a base word. For example, in English, the plural "'s" is a suffix; schematically, "cat's" comprises a base, "cat," and the suffix "'s," making more of the base. But that is not the word "onlyness." Instead, "onlyness" is a *circumflex*, which requires both parts to fully form the word; you *joined with others* are powerful enough to turn even "wild" ideas into new realities.

ZACH WAHLS AND WHAT MATTERS

Brand and onlyness also have different objectives. The central challenge in branding is to define *your value*; the central challenge in onlyness is *relational*—to meaningfully connect with your fellow denters, which is key to the scaling of your ideas in the networked age. The following story of four young boys illustrates how much relationships matter.

By the time that Zach Wahls was about ten years old, he knew what it was to lie about himself. When another kid asked, "What do your mom and dad do?" his straightforward answer often prompted teasing and taunting, as Zach's parents are a same-sex couple. After the second and third time this happened, he learned to obfuscate, or change the topic, or, if need be, to out-and-out lie.

"I'm not gay," he explains, "but I know how it feels to be in the closet."

Zach acknowledged what it meant to deny himself, his story, his onlyness. But in order to make his dent, he had not only to come to terms with himself but to convince others to join with him. He, along with many others, ultimately convinced the Boy Scouts of America (BSA) to change its policies to become more inclusive.

Zach grew up as a Scout; he can rattle off the Scout values faster than I can order a double cappuccino. For nearly eleven years, weekends for him and his family included some form of scouting activity, whether "sewing patches on orange T-shirts, searching for raccoon and deer tracks during outdoor hikes, selling caramel popcorn after church, sanding or painting pinewood cars for the Derby Race, and of course, doing service projects in the community." His love for the Scouts, though, was challenged by what they asked of him: silence.

Zach is the child of two lesbians: Terry Wahls, an internal medicine physician, and Jackie Reger, a nurse. He's one of two million children of same-sex parents born in the United States, a group that occupies a rapidly growing place within the LGBT community. Zach, like many other children in these families, learned to hide who he was. Little did he know when he was growing up that being a gay Boy Scout leader was not allowed. While there had long been many closeted gay Scouts and Scout leaders, being open about it meant getting kicked out of the organization, for it was viewed as "promoting the politics of the gay agenda." The BSA's official policy amounted to "don't ask, don't tell" for kids and their troop leaders.

Zach is not alone in having had to hide some part of himself. Christie Smith of Deloitte Consulting conducted research that found that 61 percent of people do not reveal their true selves. In order to "fit in," the majority of us admit to "covering"—trying to conform to the mainstream—even if it means not being who we actually are. Some typical examples are a gay man hiding his sexuality with "manly" sports talk, a younger man wearing glasses he doesn't need in order to appear more experienced, or a woman striving to appear tough in a predominantly male environment. But it's not just such familiar "outsiders" who admit to high rates of covering.* A whopping 45 percent of straight white men report covering as well, playing to a role or an archetype instead of feeling free to be themselves.

"Fitting in is the greatest barrier to belonging," Dr. Brené Brown, the vulnerability researcher, has written. Fitting in means adjusting yourself to meet the expectations of others. When people repress themselves, they are likely repressing their ideas as well, which suggests that some sizable percent of potential creativity is simply lost.

Suppressing oneself affects not only the individuals involved but the very fabric of our society. It obviously hobbles the wealth of ideas,

---

* Christie Smith's research found that 83 percent of LGBT people cover, as well as 81 percent of people with disabilities, 79 percent of blacks, 67 percent of women of color, 66 percent of women, 63 percent of Hispanics, 61 percent of Asians, and 45 percent of straight white men. The full report (written with Kenji Yoshino), *Uncovering Talent: A New Model of Inclusion* (Deloitte University Leadership Center for Inclusion, December 6, 2013), can be found here: http://www2.deloitte.com/content/dam/Deloitte/us/Documents /about-deloitte/us-inclusion-uncovering-talent-paper.pdf.

which in turn limits growth and innovation. But it also affects the future of meaningful work in a world where many jobs are filled by increasingly smarter artificial intelligence. For example, if men think their jobs need to be "manly," they may not seek out less "masculine" fields, such as well-paying nursing jobs. Their marginalization (by themselves and by others) reflects the economic and societal cost of fitting in.

When Zach received his Eagle Scout award—an achievement only 5 percent of the kids who start out as Scouts earn[9]—the language for the ceremony was rewritten to reflect the reality of his family of two mothers because "everything my moms have done to support me, they each needed to be included." While Zach's own troop leader found a way to ignore the official policies, Zach was always conscious of the fact that that policy existed. His organization stressed the value of honesty and honor, yet by its actions it fostered deception and assimilation.

That lesson was one he learned not only from the Scouts. In school, he was given a homework assignment to watch the Republican National Convention. This was 2004, when Senator Rick Santorum gave an address warning how dangerous gay marriage was. Watching the speech at the age of thirteen, Zach was shaken. It was the first time that he realized that "my very existence was political." When a person in authority tells you that your existence is unacceptable, it's hard to own your power. Notice that the message isn't just that "you're different" or that "your behavior is wrong," but that your very *self* is "wrong." Identity politics—when people of a particular religion, race, social background, etc., form a political bloc—can be a way for a dominant cultural group to designate almost anyone as persona non grata in society.

The following day, the teacher asked Zach to report on the convention speeches before the entire class. But he couldn't say anything, for, as he recalls, "I was too scared . . . This guy is a United States senator, and I'm just a young kid in Iowa, and there was no way my opinion mattered as much as his."

At the same time, his experience of watching the convention speeches

ultimately had an impact on his view of the world. If he continued to
remain silent, he knew he "would be saying that Santorum was right."
Less than a year later, he had the opportunity to find his voice. During
his freshman year of high school, kids would commonly use the word
"fag" in disrespectful ways around him. It bugged him so much that he
finally decided to say something, not to one person but to the entire
student body. Zach wrote a guest column in the school paper in which
he told his classmates: "So, do me a favor. The next time you hear somebody say, 'Look at that fag,' or 'What a queer,' ask them to think about
what they just said. Maybe we'll attend a school where you are defined
by your character and not by your sexual orientation. But, without a lot
of self-restraint, I sincerely doubt it. A guy can dream."

"It was a coming-out of sorts," says Zach. "I was, in effect, in the
closet before writing this piece. Being in the closet comes with a bunch
of baggage: a certain amount of self-loathing, the doubt of whether
people will still like you. The closet is just an awful, awful place to be.
We shouldn't force people to be in it."

Buoyed by the positive response from the column, three years later
Zach chose to be "outed" before the entire state. In April of 2009, an
Iowa Supreme Court decision legalized same-sex marriage in the state.
Zach wrote another piece for the school paper, celebrating the advent
of marriage equality. His high school journalism advisor recom-
mended that the piece be reprinted by the *Des Moines Register*, the paper with the largest statewide readership. With little hesitation, Zach
agreed.

Readers responded by sending a voodoo doll to his house, but also
letters of admiration and appreciation.

In 2011, when Zach was nineteen and a sophomore civil engineering
student at the University of Iowa, an organization that had noticed
his piece in the *Des Moines Register* asked him if he would testify before the Republican-controlled Iowa House of Representatives about a
proposed constitutional amendment to ban gay marriage. He drafted
some comments over the weekend, put on his one "Sunday best" suit,

and went to testify. In a brief three-minute address, he told the assembly how his mothers had raised him with good moral character, and with love, and that this love had nothing to do with the fact that they were gay.

He went on to make an eloquent argument for equality.

> So what you're voting here isn't to change us. It's not to change our families, it's to change how the law views us; how the law treats us. You are voting for the first time in the history of our state to codify discrimination into our constitution, a constitution that but for the proposed amendment is the least amended constitution in the United States of America.
>
> You are telling Iowans that some among you are second-class citizens who do not have the right to marry the person you love.
>
> So will this vote affect my family? Will it affect yours?
>
> In the next two hours I'm sure we're going to hear plenty of testimony about how damaging having gay parents is on kids. But in my 19 years, not once have I ever been confronted by an individual who realized independently that I was raised by a gay couple. And you know why? Because the sexual orientation of my parents has had zero effect on the content of my character.[10]

"It was a love letter to my parents," Zach says of that speech. He also felt a sense of duty to stand for others who couldn't speak for themselves. "Once an Eagle Scout, always an Eagle Scout," Zach explains.

A day after his address, a colleague at the University of Iowa's IT department gestured him over to her desk to see a video on Facebook. It was his talk, he realized, which had been posted online without his knowledge. Three days later it reached a million views. The *Economist* reported on his address, saying "Zach won the argument . . . even if he didn't affect the law."[11]

That speech later got him invited to address the 2012 Democratic National Convention. His prime-time speech to the 23,000 people gathered in the auditorium was televised and cemented his position as a national leader on the topic of gay rights at the age of twenty-one.

At around the same time, even though Zach was no longer a Scout, he noticed that the BSA had reaffirmed its policy prohibiting "known or avowed" gay Scouts and Scout leaders from participating in the organization.

Embracing his hope for equal rights for gays, he decided it was time to get the BSA to change its policy. To do so, he and several of his friends formed a 501(c)3 organization, Scouts for Equality. Zach created online petitions to address the BSA's donors, because major corporations might not realize they were supporting a group with antigay policies. He gathered 31,000 signatures to convince Intel to stop funding BSA (a total of $700,000), and 83,000 signatures to persuade UPS to withdraw its funding ($150,000).

Despite losing nearly $900,000 in corporate support, the BSA continued its discriminatory policies. While Zach had achieved some major victories, he wasn't yet able to make the significant dent that was his goal. To do so, he needed something more—more than just the scale that technology provided to reach strangers on the Internet.

WHAT'S NOW POSSIBLE

We often treat social media technology as though it itself is the source of scale, as if somehow, something automagically happens online that transforms the "individual you" into a more powerful "connected you."*

---

* Lee Rainie and Barry Wellman's book, *Networked* (Cambridge, MA: MIT Press, 2012), makes the argument that our society is now shaped by "networked individualism," which they stress is not isolating or isolated but rather a world according to the connected

While technology does play an important role, it's simply an *enabling* one. What actually creates scale is what binds us—the purpose and meaning and shared values we have by virtue of onlyness. An individual without clear purpose can use a website to reach others, but no community will emerge from the effort to act as one and create the necessary scale to make a dent.

The ability to find one another through social media based on shared interests, meaning, and passion is unique to our time. We can now connect to one another not on the basis of a shared heritage—where we were born, the color of our skin, our local neighborhood, a shared gender, or what school we went to, for example—but rather on a shared perspective. Andrew Solomon has described this ability to connect to those with common interests as a "horizontal identity,"[12] which he defines as an inherent or acquired trait that is generated through *choices*. He compares this to the traits people get from their parents, or "vertical identities." Race is a good example of a vertical identity, as it is passed down through DNA, as is language, because those who speak Greek, for example, raise their children to speak it as well. Solomon cites genius, autism, and sexual orientation as three traits that are horizontal in nature, because they aren't transmitted via DNA or through shared cultural norms within the family.

Central to this larger perspective on identity is the notion that connectedness and belonging are not just affiliations between yourself and others but entail a fundamental difference in the way someone conceives of himself.* This expanded sense of who you are also expands

---

me. This book review is a great synopsis: Jen Schradie, review of *Networked: The New Social Operating System*, by Lee Rainie and Barry Wellman, *Social Forces* 93, no. 4 (March 2016), https://academic.oup.com/sf/article-abstract/94/3/e89/2461413/Networked-The-New -Social-Operating-System-By-Lee?redirectedFrom=PDF.

* Marilynn Brewer and Wendi Gardner of Ohio State University researched this question of "Who is this we?" to understand how the sense of social self has evolved. Their research is unusually readable; see Marilynn B. Brewer and Wendi Gardner, "Who Is This 'We'?: Levels of Collective Identity and Self-Representations," *Journal of Personality and Social Psychology* 71, no. 1 (July 1996): 83–93, http://psycnet.apa.org/index.cfm?fa=buy .optionToBuy&id=1996-01782-006.

how you connect, from where you *came from* to what you *care about*. Onlyness is driven by this "horizontal" perspective on purpose: moms who want their child to feel okay about being geeky, people wanting fresher foods made available to them, victims of amyotrophic lateral sclerosis (ALS) wanting to cure the disease, and so on. This purposeful connection has existed for centuries before the advent of the Internet, of course, but today you can find people who care about what you care about easily.

Zach didn't realize it at first, but he was not a lone voice fighting for equality for gays in the Scouts. In parallel with his efforts, others were working to make *their* dent.

Like Zach, Ryan Andresen grew up loving the Scouts. From the time he was seven, he was mesmerized by the older kids who earned their Eagle Scout rank. He would ask question after question about what projects the recipients had done to earn the necessary merit badges and what they had chosen for their big service project.

For years, he worked diligently to earn his own merit badges, and for his service project he created a mural at the local middle school, an effort that took months. The final step in earning his Eagle Scout rank required Ryan's Scout leader to sign off on the official paperwork, so he tried to arrange a meeting to get the document completed. But the scoutmaster refused. When Ryan's father, Eric, interceded, the scoutmaster explained the situation: "The Eagle rank application from Ryan Andresen of Moraga, California, wasn't approved because of 'membership standards,' specifically 'duty to God, [and Ryan's] avowed homosexuality.'"[13]

Like Zach, Ryan had to confront the question of who defined who he was and determine what mattered to him, to decide if his cause was worth fighting for. Would he try to fit in to the world as it was or change it to include people like him?

"The very reason I loved the Scouts—honor—was what I felt I was being challenged to give up. I couldn't accept that. I was taught in the Scouts to tell the truth. Why should I be rejected for being who I am and being honest about it?"

Ryan's parents reached out to Scouts for Equality via Facebook, and Zach and his friends directed them to Change.org, where they posted an online petition to solicit support for their cause.[14] Karen, Ryan's mother, got involved because petitioners have to be at least eighteen. She wrote an eloquent plea for support, sharing how Ryan, assisted by elementary school students, had built a "tolerance wall" at his middle school. Containing 288 unique tiles, the wall illustrates that victims of bullying aren't alone and that bullying hurts real people. "Ryan has worked for nearly twelve years to become an Eagle Scout, and nothing would make him more proud than earning that well-deserved distinction," Karen wrote.

Ryan originally hoped that with his own and his parents' friends he could get a hundred signatures on the petition, which would prove to him, "I'm not completely alone." Those hundred signatures would also be sent to his scoutmaster, to convince him that change was needed.

In fact, the petition soon reached 100,000 signatures. His story was reported on Fox News, but not in a supportive way. Then Ellen DeGeneres and Anderson Cooper—two popular television hosts who both happen to be gay—invited Ryan to be interviewed on their shows.* Ultimately, 479,000 people joined his cause.[15]

It initially appeared that Ryan had won his battle, as the local troop and related council vowed to challenge official policy. Again, however, BSA refused to alter course.

But even more people were working on the same dent.

Months before Ryan appeared on television, two brothers from Chevy Chase, Maryland, had also committed to changing the BSA policy. Lucien, the elder, had circulated his own petition and gotten more than 128,000 signatures so that his brother, Pascal, could earn his Eagle rank. Pascal was ready to apply and, with Lucien's coaching, to speak out against the discriminatory BSA policy.

---

* *The Ellen DeGeneres Show* was what brought this story to my attention. "A Boy Scout Without a Badge," *The Ellen DeGeneres Show*, YouTube video, 4:52, October 10, 2012, https://www.youtube.com/watch?v=IGrz6TJk3dA.

It was this cumulative effort of this group of boys—who all shared a common purpose, born of their own individual meaning—that finally created the dent.

In May 2013, the Boy Scouts of America's national council voted to end the ban on gay Scouts. The new policy took effect January 1, 2014. The following month, Pascal Tessier became the first publicly gay Scout to be awarded the rank of Eagle Scout. In late May 2015, the president of the BSA called for the end of the ban on gay Scout leaders.

These boys could have easily listened to those who told them that they were wrong for being who they were and characterized their ideas for the world as "too wild." But in pursuing that thing that only they saw, they found one another and were able to create real and lasting change. They were working to change BSA policies to benefit not just themselves but thousands of other boys, boys who deserved to have a chance to make a difference despite and independent of their sexual orientation.

DON'T BE A LONELY ONLY

Community is central to onlyness, for it enables you to progress from being the "only one" to enlisting the strength and scale of a group. Research from sociology, psychology, and anthropology has consistently shown that when individuals are in the position of being the "only one" in a group with a different norm, they will be pressured to conform to it. Nonconformists are typically dismissed and made to feel "other."* In those circumstances, "others" can be set up to compete with, to judge, or even to betray one another in order to win one of the limited number of seats at the table.† In many respects, conformity

---

* Otherness has deep roots. First, it was used for survival; "we" banded together against the "other" tribe. Today, being designated "the other" is often a way in which someone who is nonwhite and nonmale is made to feel as if she doesn't belong to society's "majority." It affects a great many people. Today, the United States is demographically only 31 percent white and male, suggesting that "otherness" is actually the majority, not the minority, point of view (and also suggesting that it's an outdated construct).

† The actress Kerry Washington gave a compelling speech in 2015 on this topic of people being set up to compete against one another, which I blogged about at the time: "The

or competition is not a choice but simply a matter of survival. To crack the code of acceptance and belonging—the most fundamental of human psychological needs—you'll naturally suppress the qualities that set you apart. Without belonging, the odds are good that you will give up your own ideas. As a result, your fresh perspective and novel ideas are either deferred, in the best of cases, or extinguished, in the worst.

That dynamic changes when you find people who share a common cause.

Sharing a Common Purpose

Zach, Ryan, Pascal, and Lucien all felt alone until they found the support of others who shared a common purpose, those who championed inclusiveness in the Boy Scouts. They each felt the pressure to fit in rather than be themselves until they had shared purpose.

Thus, onlyness isn't a path to loneliness; it's anything but.

Only Norm Is Onlyness," *Yes and Know* (blog), March 27, 2015, accessed February 17, 2017, http://nilofermerchant.com/2015/03/27/the-only-norm-is-onlyness.

It's as if there is a metallic thread*—perhaps catching light, and thus your attention—that only you see, and, by pulling on it, you find yourself connected to a larger fabric of society. You finally find the way to be deeply attached to the world, not by fitting in but by standing by your own ideas.

People who defy convention are labeled with many disparaging terms: rebel, freak, square peg in a round hole, weirdo, black sheep, oddball. They are pigeonholed as "misfits" until they find those who share their goals and passions. The dynamic changes at a point of critical mass; research has shown that at least 30 percent† of a group has to consist of nonconformists before the "other" label is abandoned and each member is valued for him- or herself. When there are *enough* people valued for their ideas rather than the power of who brings those ideas, the best idea has a shot.

THE CLASH OF IDEAS

Zach Wahls has continued to make Scouts for Equality his main effort, and along with those who have united behind the project, he has made huge progress. He describes his work as fundamentally confronting a clash of ideas, a debate about who is allowed to count:

> Zoom in as closely as you want, or zoom out as far back as you want. You can make this about gay inclusion, gay marriage, LGBT rights or human rights. But the issues are the same. One group wants to impose a strict worldview upon themselves and others of what is right. The second embraces reason, and progress,

---

* I believe this thread analogy was something I read in poetry by Rumi, but for the life of me, I can't find the original reference.

† This statistic comes from Beate Elstand and Gro Ladegard, "Women on Corporate Boards: Key Influencers or Tokens?" *Journal of Management and Governance* 16, no. 4 (November 2012): 595–615 (first online: November 20, 2010), https://papers.ssrn.com/sol3/papers.cfm?abstract_id=1582368.

and believes in the freedom of everyone to count, to
matter, to contribute. This fight has always been about
how and when different people are allowed to be at the
table as equals.[16]

   As the stories of Kimberly Bryant and the Scouts who are advocating for equality have demonstrated, onlyness is how seemingly wild
ideas come into the world and are embraced by those who care. Ryan
Andresen once eloquently expressed his goal as being to "reach beyond my ego and my puny little self" to make a groundbreaking and
lasting difference. Rooted in meaning and connected in purpose, anyone now has access to this power. Even you.

CHAPTER 3

Discovering Yours

> There is a vitality, a life force, an energy, a quickening that is translated through you into action, and because there is only one of you in all of time, this expression is unique. And if you block it, it will never exist through any other medium and it will be lost. The world will not have it.
>
> **—MARTHA GRAHAM**

ANDRÉ DELBECQ: BRING ALL OF YOU TO WORK

"We are God's stake in human history. You are each called by name."

Listening to this statement are evening MBA students in a fluorescent-lit lecture hall at Santa Clara University. Feet shuffle and eyes turn to mobile devices to avoid the professor's gaze. Amid this rainbow coalition of people—Christian, Jew, Hindu, Muslim, atheist, and even agnostic—all seem equally uncomfortable.

André Delbecq is no longer startled to find himself leading this class, though the very idea of teaching how faith could coexist with fortune was once one he could never have conceived of. Yet, this is his dent.

André's discovery of his purpose sheds light on how each of us can do the same. It answers the question so many of us have as we look over the edge to a possible new world: Don't we need to know where we're going and, ideally, have a well-considered plan before we start? When we imagine doing the next big thing, whether it's a business innovation or a social change, most of us expect to have a transformational moment, sufficient expertise, or some brilliant introspection that lends us the clarity to act. Yet, as André's story reveals, the journey that leads you to yourself doesn't require knowing exactly where you're going or a map outlining precisely how to get there.

THE QUESTION

André's résumé was built on forty years of scholarly research in decision making and organizational design. He was a sought-after professor on the topic of innovation for rapidly changing environments. Then, one day, his colleague Elmer Burack, a professor visiting from Illinois Institute of Technology, asked him a question. As they sat on André's back deck in Alameda, overlooking his thirty-three-foot pilot-house ketch sailboat, Elmer wondered, "Now that you've done everything you imagined, with your reputation established, what's the one thing you can do now to still make a difference?"

With his retirement age only a few years ahead, André had imagined he'd be spending more time with his wife, occupied with his boat, or putting on leathers instead of suits and going for long rides on his beloved Harley-Davidson motorcycle. But he had no answer to Elmer's provocative question.

"I'll give you a book that has changed my life," Elmer proposed. As he read the book, Zalman Schachter-Shalomi's *From Age-ing to Sage-ing*, André started to ponder what it meant to "no longer be the knight" fighting daily battles but to serve in a wisdom role to help others. And one subject came to mind as he pondered what he could do as his last hurrah: "There was one question I was never able to answer for executive leaders. They would ask me, 'Do you deal with the inner life of the leader?'" André would answer, "Well, no, the management program teaches something close, but not quite the thing. Ethics is the closest a business school gets to the topic of values and meaning."

André had always found dealing with such subjects rather frightening. Even at his Jesuit university, within the management discipline they were considered untouchable, as they could not be associated with the scientific method or achieve measurable outcomes.

It's when you encounter such questions—ones that you can't answer but can't stop thinking about—that you also have the first clue to discovering what matters to you.

Which is not to say you'll always accept the question. If your contemporaries dismiss it as too wild to be worthy, it's easier to suppress your own sense of what matters, your "only." André's initial response was not unusual: He decided to get someone *else* to do it—someone who was much better suited to take on the challenge, someone "right."

WE SHOULD HIRE SOMEBODY

We should hire a Jesuit to teach that subject, André thought.

On his next trip to Rome he met with a prominent figure in the order and made his case. The cleric listened to André's argument thoughtfully and then replied, "I have often found in my career that when people come to me with something and say, 'You should do it,' they themselves are being called to take it on."

André resisted the call. "At age sixty-two or sixty-three, I know that I'm a wonderful sinner, and I figured I would resolve my brokenness before I died. [But] I couldn't define spirituality, let alone talk to it. So teaching it was impossible. I had no training in it . . . I'm the wrong guy."

Many people regard themselves as unequal to pursuing a demanding new direction when confronted with it. But a perceived flaw—e.g., being a "wonderful sinner"—may actually be what best qualifies them to assume the task. To make a contribution doesn't need perfection; it needs only you.

Soon afterward, André was invited to take part in a retreat in Loyola, California to help Jesuit priests determine their future direction. As he was the former dean of the business school, as well as a member of several corporate boards, this was a usual request. At the retreat, another question provoked him.

A sermon to guide the participants' decisions was delivered, centering on a familiar New Testament story: Peter is called by Jesus to step out of a boat in the middle of a body of water and come to him. This account is a parable about an act of faith, about taking action when any logical mind could find many a reason not to.

While the group peacefully meditated on its message, André wrestled with the question of his own inaction. *What am I even doing thinking of this possibility?* he asked himself. *I don't even know how to pray, let alone what I should do to answer questions regarding spirituality that leaders pose to me.*

His forty years of concrete* experience in teaching management, though, convinced him that the subject was important, for ethics alone aren't a sufficient moral foundation. "Hubris and greed are two major traps in senior business leadership," he explains. "Everybody knows what's ethical—what is right and what is wrong—but it's easy to be seduced by power and wealth. Unless you have a well-developed spiritual compass, power corrupts."

Long before Elmer had goaded him, he had sensed what was missing in management training and why its absence had led to such massive corruption in business today. But he never considered acting on it. The idea of doing so wasn't so much new, then, as one that had been previously denied.

---

* Manuel Castells, the Spanish sociologist most known for his research on network power, argues that ideas matter and scale in the network era when they derive from people's *concrete* experience.

A month after the Loyola retreat, André and his wife, Mili, were on their way to a meeting when Mili reminded him that they had to rent a house for an upcoming sabbatical in Bordeaux, where André would be studying innovation in the wine industry.

Conflicted, he blurted out, "I'm not sure . . . that I'm going to France." He then heard himself tell his nonplussed wife something he'd never have imagined himself saying: "I think I want to go to the graduate school of theology at UC Berkeley to study spirituality."

New directions emerge from these kinds of unexpected inklings, questions, and hints. They're what author Steven Johnson describes as a "slow hunch" about what's needed.*

TURNED BACKS

Rejection, or the fear of it, is a key impediment to many people's laying claim to their own ideas.† Will others around you "get it" or think that you've lost your mind? Will your professional colleagues judge you for chasing a question they find unworthy? Will your close circle support you even though you have no idea what you're doing?

André attended an Academy of Management meeting, where he had been the dean of fellows, the highest honor one can receive for scholarship in management. When he began to discuss his new explorations with one of the fellows, "the other man stopped the conversation

---

* Novel ideas develop through networks, as Steven Johnson suggests in his book *Where Good Ideas Come From: The Natural History of Innovation* (New York: Riverhead, 2010). At its most basic level, an innovative idea is simply a network of cells firing inside your brain in an organized pattern, a "hunch." But for that hunch to blossom into something more substantial, it has to connect to other ideas—which is where action comes in, as it helps you find other people like you. Just as thought is generated by a network of cells, innovation is built out of networks of connected ideas, and it comes to life through networks of people.

† A growing body of research confirms people are silenced by fear at work. Fifty percent of respondents in a 40,000-person survey reported that they were unable to raise important issues due to fear. See Jennifer Kish Gephart, James R. Detert, Linda K. Trevino, and Amy C. Edmondson, "Silenced by Fear: The Nature, Sources, and Consequences of Fear at Work," *Research in Organizational Behavior* 29 (2009): 163–93, http://www.hbs.edu/faculty/Pages/item.aspx?num=36259.

midsentence, turned his back on me, and walked away." He faced the silent moment awkwardly, not knowing what to do. When he saw the man returning in his direction, he was notably relieved—"until I noticed that he was not making eye contact but coming back to pick up his table tent card, which was next to mine . . . he moved it to another table, permanently."

André went from being the top dog to being rejected. "A number of colleagues distanced themselves from me the moment I went on this journey, and then when I returned, they thought I was a pariah, some nutcase." He became so despondent that at one point he almost quit the university.

But André eventually did manage to make the idea of spirituality in business acceptable within the academy: first, by being willing to be foolish; second, by acting without evidence that he would succeed; and third, by enlisting help from some crucial new allies.

GOING BLAH, BLAH, BLAH

When André decided to study theology, he was uncertain how to proceed. "I knew I needed to study, but I had no idea what to study, or what courses to take, or how to get guidance."

One of André's colleagues, Mark Ravizza, counseled him to meet with his fellow Jesuit, Frank Houdek, for spiritual direction and to do a "supervised readings approach," but André was unable to reach him. He sought out other spiritual directors, but they weren't interested in advising someone who wasn't planning on becoming a priest, especially someone who was interested in helping business executives. Faith and fortune don't mix; both sides of the equation confirmed this was a "wild" idea. The resistance he encountered only confirmed André's worst fears. He describes this period as one of "knocking on doors and either hearing crickets" or having the doors slammed in his face.

When André encountered Mark Ravizza again at a social event at the Jesuit School of Theology in Berkeley and told him of his failed efforts, Ravizza led him across the campus to the elusive Frank Houdek. André remembers the meeting vividly. "Frank was sitting in an unlit office, three walls groaning with books and a shaft of sunlight shining down on the desk. I was politely invited in. He just looked at me with a blank face while I went on and on—quite frankly, I went blah, blah, blah for a long time—about what I wanted to do and why it was important and why I felt it was a deep need for leaders. And that I had no idea how to begin."

André anticipated another rejection, but Frank's answer surprised him: "I have discerned that I will direct you. What you are doing is important, and I will meet with you every other week. And you will take a topic that you think is important, and you will write an essay, and I will respond to that essay."

The actual process proved to be slow, uncomfortable, and unclear, much like the process of discovering one's onlyness. There were no readings and no clear assignments, which was not what André, as a business professor, was accustomed to. But he persevered, though his ideas were not well received. "[Frank] would take my ideas and things that I thought were clear . . . he would put such a new twist on something that I felt like I was now seeing the idea through a kaleidoscope." Every notion André put forward had to be reworked when Frank showed him a new dimension or hidden depth to the idea.

"Was it hard?" André asks. "Yes, but also no. I had read this Orthodox notion of people who want to serve the world must be willing to be 'God's fool,' meaning they may not look the part, they may not be as ready as they need to be, but they are willing to do it anyway."

BEING THE FOOL

What does being "foolish" mean in the context of one's only?

As adults, most of us learn new things that we can place within the context of large knowledge sets that we already possess. But when we encounter something totally unfamiliar, learning is very different. The process is much like trying to learn a new language as an adult; during the first few months of study, you sound and feel as tentative as a two-year-old. In the same way, discovering purpose is exploratory and therefore leaves you especially vulnerable. The process requires a growth mind-set, as Carol Dweck, the noted Stanford psychologist, describes it.*

André's use of "foolishness" reminds me of how Jim March, a Stanford professor who studies decision making, employs the same word to explain how new ideas are formulated:

> Someone in economics, for example, may borrow ideas from evolutionary biology, imagining that the ideas might be relevant to evolutionary economics. But there is no guarantee this works. But, this kind of cross-disciplinary stealing can be very rich and productive. It's a tricky thing, because foolishness is usually that— foolishness. It can push you to be very creative, but also uselessly creative. The chance that someone who knows no physics will be usefully creative in [applying] physics must be so close to zero as to be indistinguishable from it. Yet, big jumps are likely to come in the form of foolishness that, against long odds, turns out to be valuable. So there's a nice tension between how much foolishness is good for knowledge, and how much knowledge is good for foolishness.[1]

---

* Carol Dweck's book *Mindset: The New Psychology of Success* (New York: Random House, 2006) introduced the idea of two frameworks. One is a fixed mind-set, in which someone is optimized for mastery; the other is growth focused, which is optimized toward learning. I once interviewed Carol about her ideas and asked her what was the underlying message that a person who is growth focused communicates to himself. She replied that it came down to whether someone trusted himself to rebound when he failed. You can read our interview on my blog: "Do You Trust in Your Ability to Grow?" *Yes and Know* (blog), September 27, 2013, accessed February 20, 2017, http://nilofermerchant.com/2013/09/27/do-you-trust-in-your-ability-to-grow.

André commented on his own experience, saying, "I wasn't sure where this was all going . . . and everything that happened that led to its success would look random to any outsider. Like the phone ringing, or running into an old friend, or a question raised at a meeting. But it wasn't that random if you consider that I was willing to look at it, to be changed by it. You have to be willing, and open to the new question, for all of this creative space to exist."

For your idea to even have a chance, you have to give it room to grow. In spite of whatever indications suggest that an idea might not work out, you have to be "foolish" enough to make the attempt to make it successful. You have to be willing to learn that which you don't know, to ask questions that haven't been asked before, to not know something for long enough to learn something new, and to trust that, even if you don't learn it right away, you can rebound from that failure.

This is how discovering your onlyness takes place: Give yourself permission to pursue the question you have, even if others might view it as a bit nuts. Then, do the necessary work, even if you don't know exactly how it will turn out. It's in the *process* of doing that work that you become ready to make a dent.

To design his first prototype course, André invited feedback from the people he had met along his way. Rejection by his existing community had led him to seek out new colleagues and cohorts who "got" the idea. Scholars from every faith agreed to advise him and help structure the new course.

NOT NAVEL-GAZING

In retrospect, it's relatively easy to connect the dots to trace how André went from sitting on the dock by the bay with Elmer to studying theology to prototyping a course on spirituality in business. But to what degree was his path planned or even clear, and how much was discovered in the course of his journey?

"You start out with an idealized sense of how you'll pursue a new direction, that you'll be able to map it out, and have clarity," André explained. "[But] when you have lived it, you see [the path] is not neat, nor linear, or even especially known to you when you start. There's rarely certitude or confidence as you set out. It unfolds."

He describes "unfolding" as "little agreements done with great love. By a phone call you were joyous to say yes to, or when you participate in a meeting because you are asked to do so and you see yourself adding your bit. It's how you presented a paper or an analysis to a small group with your full heart, and colleagues understood what you cared about, and this becomes your [new] reputation. And, as a result, you get into a conversation and then later introduced to someone who turns out to be a crucial piece of the puzzle." The pattern of how André discovered his path is an illustration of what you have to offer being put to use in the world. It's not planned, and it's certainly not about being packaged; rather, it's about being yourself in the moment so you contribute what only you can to others, to add value in a way that serves others.

It's worth noting that André did not spend much time in navel-gazing. One idea that is deeply engrained in Western culture is that introspection (aided by self-help books or self-assessments) can lead to clear action steps and maybe even a better outcome for a given idea. Sometimes such resources can help with personal clarity, but introspection can just as easily keep you from moving forward. Herminia Ibarra, a leadership professor whose expertise is in navigating career change, urges, "Devote the greater part of time and energy to action, because that's when you discover what you care about, [and identify and] meet the people who can help you, to form the [basis] to be successful in the new thing."[2] After all, if you navel-gaze at the beginning, you won't have any new information to parse, and your existing tribe will likely tell you, "That idea is *too* weird, *too* wild." Because for them, it is.

The capacity to act without assurance of success is an incredibly important factor, as Kimberly Bryant demonstrated in her launch of Black Girls Code and the Scouts in their campaign for equality. Research shows that powerful people actually take *more* action than those

less powerful.* Modern research on *interpersonal* power reveals that
your agency—your ability to act—is one key to being more powerful.†
Put it all together, and it's clear: While navel-gazing may provide tem-
porary comfort, it's action that generates change, both in you and in
what you want to achieve.

André's story illustrates that discovering what matters to you and
how you want to address it starts with a question.

Some people have many topics that are worthy of their attention,
and they will be most effective by focusing on one. You might be
served by the words of author Cheryl Strayed: "Stop asking yourself
what you want, what you desire, what interests you. Ask yourself instead: *What has been given to me?* Ask: *What do I have to give back?*"³
Then, act.

What about those of you who don't *have* a question? To be sure,
there are some who can't immediately identify something they passionately believe in. It could be that your dent right now isn't a thing to
do, but a way to *be*, perhaps to find onlyness in people around you and
help them to make their dent.

And maybe some of you hope to find the "perfect" dent—as if there

---

* In one experiment, people were primed to feel powerful, and it turned out that they took
action; see Adam D. Galinsky, Joe C. Magee, Deborah H. Gruenfeld, and Jennifer A.
Whitson, "Power Reduces the Press of the Situation: Implications for Creativity, Conformity, and Dissonance," *Journal of Personality and Social Psychology* 95, no. 6 (2008): 1450–66,
http://wagner.nyu.edu/files/faculty/publications/mageePowerPressSituation.pdf. Three
researchers from the paper just cited (Galinsky, Gruenfeld, and Magee) argue that power
is not just a structural variable (like money or title or whatever) but also the psychological property of an individual. It turns out the very act of doing *something* activates cognitive power. So, while power and action have long been associated, the thesis has always
been, very specifically, that power leads to action. But it turns out the opposite is true, too:
Act, and you are powerful. See also Adam D. Galinsky, Deborah H. Gruenfeld, and Joe
C. Magee, "From Power to Action," *Journal of Personality and Social Psychology* 85, no. 3
(September 2003): 453–66, http://psycnet.apa.org/journals/psp/85/3/453.
† Researchers Rachel Sturm and John Antonakis did a review of the academic literature of
power; much if not most of it is tied to organizational context and thus less individuated.
But it is a solid overview on the relevant literature on power; see Rachel E. Sturm and
John Antonakis, "Interpersonal Power: A Review, Critique, and Research Agenda," *Jour-
nal of Management* 41, no. 1 (January 2015): 136–63 (first published: October 24, 2014),
http://journals.sagepub.com/doi/pdf/10.1177/0149206314555769.

is only one thing that matters, and it will become clear one day, out in the future. If you're one of those people, put this book down, rub your eyes as if you're waking from a slumber, and look around yourself, devoting your full attention to your surroundings. You might notice a neighborhood kid who needs help with math, or a homeless man who lives under a nearby bridge, or a government policy that you disapprove of. Instead of waiting for some "perfect" way to add value, maybe just see what needs to be done.

Even if that action is small.

When Zach first began his campaign, gay marriage was illegal in most of the United States, but that didn't stop him from speaking up as an advocate for a world that included his family structure. His first step was calling out how it hurt others to hear words like "gay" and "fag" used derogatorily. From there, he proceeded to argue why equal marriage rights wouldn't hurt anyone. One small act led to another. This strategy of achieving small victories is a powerful way to tackle a much larger issue. Part of the reason why small wins can be so much more effective than anticipated is how humans deal with novelty. Merely defining something as "big" and "serious" can be daunting to the point of discouraging any attempt to confront it. By doing small acts, one after another, you build sufficient momentum to tackle the ultimate goal.

HOW THE WILD BECOMES NORMAL

At Santa Clara University, André had the support of Barry Posner, then dean of the business school and one of the top leadership authors in the world. Barry encouraged André's proposal to study theology, saying, "You will be the business school's contribution to the Jesuit mission. No one else is doing anything like this. So I give you full freedom to do this research and to teach us."

André taught the first prototype of his course—Management 696, The Spirituality of Business—in the fall of 1998.[4] Barry again offered his assistance, as he had the authority to allow the class to be taught as an experiment for up to three years. During that period some five hundred MBA students (who were also working professionals) and three hundred senior business executives attended the class, which received some of the highest marks in the school, yet the Graduate Studies Program Committee—the group that approves courses—denied it status to continue. The notion that the topic of God would have a legitimate place in a business school—even a Jesuit one—was simply too extreme to be acceptable.

It would take several more years before the course was finally officially approved, but during that time, André gathered the support of allies and co-denters. A management, spirituality, and religion interest group was formed at the Academy of Management. André discovered five other professors who were engaged in efforts parallel to his own.* Elements of his curriculum were adopted at the University of Dallas and the Tyson Center for Faith and Spirituality in the Workplace at the University of Arkansas, and in parts at various leadership centers and universities around the world.

Over the decades, André had borne witness to major paradigm shifts in the teaching of business. Areas once deemed unsuitable for study at a business school had become mainstream—behavioral science and mathematical modeling in the 1960s and 1970s, corporate responsibility and ethics in the 1980s and 1990s. He had now become one of the leaders within the academy to help initiate the new field of spirituality in business.

---

* One important paper written by five professors tied the early findings in this domain of spirituality of organizational leadership to strategic decision making. See André L. Delbecq et al., "Discernment and Strategic Decision Making: Reflections for a Spirituality of Organizational Leadership," https://pdfs.semanticscholar.org/8d13/519168be23a 120b64e4680c1cbbdb598e028.pdf.

CHARLIE: MY STUDENT

Despite the value of asking a new question, many people stop short of pursuing it. One overriding reason that keeps them from chasing an idea they think is valid and interesting is the difficulty of shifting from their own success to focus on an idea beyond themselves.

"Getting accepted into the vaunted incubator [Y Combinator] seemed like a dream come true—until it became a nightmare," writes Charlie Guo,[5] a former student of mine at Stanford. Only three months after Charlie and his cofounder, Kevin Xu, had joined the start-up accelerator, they were warned by its cofounder Paul Graham, "If this was a class, I'd be the professor warning you that you're in danger of failing." In fact, I had given Charlie notice that he *was* failing only a few months earlier. Charlie's story illustrates how looking for success rather than looking at a problem more deeply can distract focus from what is truly meaningful.

HOT IDEAS ARE NOT ENOUGH

I first met Charlie in a class I co-taught (Management Science and Engineering 272—Start-up Boards, an advanced entrepreneurship course at Stanford University) with venture capitalist Clint Korver and four other lecturers. The course aimed to serve soon-to-graduate students with a chance to prototype an idea and build out a go-to-market plan, thus incubating a concept into reality.

Charlie and his pal Kevin came armed with an idea for the education marketplace, which they termed "Codeacademy on steroids." It had goodness at its core, as it proposed a way to help underserved kids learn to code. Charlie had already prototyped the idea via GitHub and could prove that, with the right types of support, anyone could learn to code. This would not only serve a big market need, but would give kids who would not otherwise have had one a shot at good jobs and a future. As an educational platform to create more opportunity for all, the idea warmed my heart, and Charlie and his partner seemed so committed, so full of conviction.

In the best of cases, every start-up idea is born of onlyness. A start-up may use a different word—like "authentic," "mission driven," or "passionate"—to describe itself, but at the root of what makes a great start-up or a great entrepreneur and team is what underlies onlyness: a different point of view grounded in one's history and experience, visions, and hopes. That's what I thought I perceived in Charlie, but as early as the first class meeting, my colleagues and I had a clue that something was amiss.

Class sessions are deep work. We ask all the questions necessary to ultimately build a viable business: Who is the customer? What is the market? Why would your customer pay money for this service? How will these sometimes underserved groups be able to prove they've gained something of value? What are the ways you can measure progress? Can you certify them so they can be qualified in the marketplace?

As the students delve more deeply into their ideas, they naturally discern flaws in their logic, learn new techniques, lock down certain facts that must be true for other things to happen, and keep refining their plan. When Charlie offered answers to instructors' questions, however, he would change whatever his first response was in the face of any skepticism. On issue after issue, he would shape-shift away from a clear and consistent perspective. He was dancing for us, not thinking. After a while, we could tell there was no through line or unifying point of view to his replies, and Clint finally told Charlie what all of us professors were thinking: "I don't know if you believe in your own idea."

I advise a lot of entrepreneurs, and the ones I worry about aren't those who are stuck on finding the best answer but, rather, the ones whose question changes so often that they wind up going in circles. In Charlie's case, there didn't appear to be a *there* there, a mission or a challenge that *had* to be taken on. He didn't seem to be driven by a genuine passion to serve the kids in his target audience. His idea could have been one of a thousand picked out of the sky simply to "do a start-up." This team clearly had skills but seemed to lack meaning.

One specific situation brought the problem home: During a deep-dive session, we were probing Charlie's demographic. His potential students might lack good study habits and not have access to peers or adults who were prepared to help them through the rigors of learning coding.* Could they be expected to accomplish this on their own? The lessons couldn't all be just technology enabled but had to offer on-demand peer help. "Kids need a social construct," I told Charlie. "It's not a lack of knowledge but a peer network that enables people to learn. I have trouble seeing this idea working without that." They didn't have a response to our challenge, so we asked them to investigate the matter further, offering ideas on how they could do so.

This assignment would obviously entail Kevin and Charlie's getting together with some of their existing pilot users (recruited from the Reddit community) and exploring unfamiliar terrain: Did their online coders feel they needed peer support? Did the ones who were succeeding with the current beta tool have existing peer support? The following week, during our status update, Charlie and Kevin gave us blank looks when we asked about what they had learned. "Well," they explained, "rather than talk to a bunch of random people, we thought we should focus our time on coding up the user interface."

Why didn't the two men want to talk to their customers? Did they fear it would be awkward? Did they not want to risk looking unprofessional? Whatever the case, they did not embrace André's openness to

---

* This line of inquiry was based on the learnings of Uri Treisman, a mathematics professor who figured out that, contrary to popular belief, his African American students were not performing poorly due to a lack of motivation or preparation but, rather, because of their social and academic isolation on campus. See Uri Treisman, "Studying Students Studying Calculus: A Look at the Lives of Minority Mathematics Students in College," *The College Mathematics Journal* 23, no. 5 (November 1992): 362–72, http://math.mit.edu/~hrm /interphase/TreismanXArticle.pdf.

being "foolish." Charlie and Kevin were playing it safe by staying on familiar ground.

Sometime during our penultimate class session, the students were invited to pitch to Y Combinator, and Charlie and Kevin were accepted, which they interpreted as validation, both professional and personal. But they neglected to realize that both Y Combinator and the class offered the *same* opportunity: to dig in harder, to do work that mattered in the most effective way possible. Charlie's confessional post about failing Y Combinator is excruciating to read, but in it he offers a perceptive insight when he acknowledges: "I had impostor syndrome."

PERPETUATING THE IMPOSTER SYNDROME

Imposter syndrome occurs when you feel inadequate and worry that others will "find out" you're a phony and don't deserve to be in the position you're in. Many people experience this at some point in time, but why does it happen? And, more to the point, is it avoidable?

The key to avoid falling prey to this feeling is where you place your focus: Is it on you, the ego-filled "I," or on the purpose, the "it"?

If your goal is based on acting out a role or merely striving for success, then you're likely to be prone to imposter syndrome. It occurs when you try to look the part of a badass entrepreneur rather than concentrating on whom and what to serve distinctly well. A trap of imposter syndrome is that it invites exactly the type of behaviors that only make it worse. Feeling like an imposter often drives people to puff themselves up, to project even more bravado. They start to tell slightly embellished stories, and bit by bit they create an illusion of someone other than themselves as they really are. Not only have they wasted time and energy with the wrong focus but they ultimately lose themselves in the distortion.

Luckily, there's an easily accessible antidote to imposter syndrome. It is simply to be yourself, and to be so deeply committed to a purpose that what matters is not how you appear to others but what is actually served. Take good care of your actions and your reputation takes care of itself. We each discover what matters not by focusing on our own needs but by paying attention to those of the world. The transformational power of just being yourself takes you outside of the "I" of self-interest to care about "it," a purpose. This shift in attention is how a dent can really begin to take shape.

So why not chase success directly? Doing so may mean that you never make your dent. It tempts you to take your eye off the ball, just like focusing on the packaging rather than the purpose.

You know what doesn't feel like imposter syndrome? When you're being yourself.

How did Charlie's story turn out? He has since left the country, working remotely as a software consultant, and I can't wait to hear what path he ultimately chooses.

Until then, though, his experience leaves us with constructive lessons. Ideas are relatively easy to come by, but conviction demands the energy of purpose. We need to know that something matters, and why. Conviction is willingness to do the work, to live with uncertainty, to be open to asking for help, and not to worry about the end result. Real confidence (as opposed to bravado) is born of committing oneself to that work.

CINDY GALLOP: CHANGING HOW WE HAVE SEX

When Cindy Gallop was in her midforties, a chance series of events offered her the opportunity to date a number of men in their twenties. Though she thoroughly enjoyed these encounters, she made a grim discovery: These twentysomethings had a distorted notion of what constituted good sex. While porn has been around forever, the ease of access to Internet porn had been subtly warping the perspectives of an entire generation and, in the process, intimacy was being lost. Astutely, she saw a potentially serious social problem developing.

That's how Cindy, by doing what she loved, quite literally, founded her second start-up: Make Love Not Porn (MLNP).

This former advertising executive who entered one of the most taboo of fields with the aim to disrupt it offers a provocative yet profound view of how to discover one's onlyness. She did so not by chasing a question, as André did, but by tackling an issue directly in front of her. More to the point, she did so by speaking her truth and valuing the power of her own ideas, even when they weren't mainstream. Noticing what is immediately before you—especially when it would be all too easy to overlook it altogether—provides us another template for how to discover what matters.

OUT OF THE SHADOWS

MLNP is designed to be an alternative to porn, which is the dominant framework for content, language, or visuals to suggest what "good sex" is. It's not anti-porn, because in Cindy's mind, that's not the issue. The dent she thinks is necessary is that we as a society talk about sex as it exists in the real world. MLNP aims to build a new category—sex presented online via a crowdsourced, user-generated, video-sharing platform that showcases and celebrates what Cindy hashtags as #real worldsex in order to normalize sex as a topic and promote conversation that leads people to understand their own sexual values. The business model involves subscribers uploading videos, paying $5 to post, and users paying $5 to watch, with 50 percent of the proceeds going to contributors.

Cindy is quick to point out that porn has a very limited narrative, one that is conveyed in mostly male-oriented language. Where porn dialogue uses terms like "'slamming,' 'thrusting,' and so on," Cindy and her team of four opt for language like "'juicy,' 'funny,' and 'hey, hey, *hey*, now.'" "Sex should be a generous act. It should not be a transaction," Cindy says. "When sex is a one-way fulfillment, or a playing of a role you're copying, you're not really there. You're not connected . . . with yourself, or in contact with the other person. Real-world sex is present and real and generous." It is also, she adds, enjoyable, playful, earnest, funny, sweet, awkward, exploratory, powerful, intimate, shy, brave, and even spiritual. "During real-world sex," she says, "you can be *all* of you. And you can show all of you, not some façade."

Cindy's goal is effectively "normalizing" real-world sex. "Normalizing"* is the word used by Shonda Rhimes to describe her characters and story writing. A prolific director and producer who is responsible for ABC's programming on Thursday nights, Shonda writes to reflect the world as it is. Her shows (*Grey's Anatomy, Private Practice, Scandal*, and others) include people of all ages, races, genders, origins, and faiths. As in, *normal*. For its part, the normalized real-world sex featured on MLNP is flabby, funny, comes in all colors, and involves all types of activities.

Cindy's hope is that her website not only leads to more intimate sex but provides an accessible way to talk about it. "The average age that kids see porn is eight," Cindy has discovered in her research. They don't necessarily go looking for it, but they can all too easily end up in

---

* You can read more on normalizing in Shonda's book, *Year of Yes: How to Dance It Out, Stand in the Sun and Be Your Own Person* (New York: Simon & Schuster, 2015), and online: Derrick Clifton, "Shonda Rhimes Just Came Up with a New Word for 'Diversity'—Let's Start Using It," Mic.com, March 17, 2015, https://mic.com/articles/112914/shonda-rhimes-thinks-we-need-a-new-word-for-diversity-here-s-why-she-s-right#.fpuCSBfR2.

the darker corners of the Internet where it resides. Society, she argues, needs to have more online resources to help everyone view sex as a healthy and normal part of life.

The next step in Cindy's plan is to add educational resources to MLNP. Sex educators could post videos, materials, and tools that schools could download for age-appropriate resources. These resources could also help parents talk to kids about sex and, of course, lovers to talk between themselves. It's a grand vision—a bold idea for a dent—but does it make Cindy an activist, or a revolutionary?

No, she answers. "I don't like labels. I am just doing what I want everyone to do: standing up for the things they believe in, the ideas that matter to them. I think that's the right thing for people, for the industry, and for our world overall. I am simply doing what is true for me, sharing beliefs, and advocating for my values, because no one else can do that but me."

DON'T CARE WHAT OTHER PEOPLE THINK

Four stories trace Cindy's journey to her *only*. The first concerns her childhood.

Cindy is the product of a mixed-race marriage. Her father is English, and her mother is Chinese. In 1959, when her parents began dating, her mom "had to crouch down in the back seat of my father's car, so the neighbors couldn't see that she was dating a white guy." Growing up, Cindy remembers hearing that when her parents got married, one member of the English side of her family was heard to say, "Pekes should stick to Pekes, and pugs should stick to pugs."* People would regularly say to Cindy, "You're a bloody half-caste, ain't ya?" "Growing up in Asia," she recalls, "I encountered racism from both sides [of the population]—to the English I wasn't English, to the Chinese I wasn't Chinese."

Being seen through the lens of any "ism"—in Cindy's case, racism,

---

* The reference was to Pekingese dogs—a Chinese breed—and pugs, an offshoot of British bulldogs.

classism, and sexism—limits who you are. Isms are useless adjectives; they are simply categorizations of the "other." Anytime we're seen through the lens of a stereotype or group, we're not being seen as ourselves, with the distinct ideas we have.

Cindy came to realize that if she "cared about what someone else thought," she'd "have been nothing." Today, if you see her on the speaking circuit, you can tell she's speaking from experience in insisting that fitting in has a cost. To be considered "good" and "a fit" is often equated with being like everyone else.[6] This means the reverse is also true: If you are not like "them," you can be dismissed as invisible, flawed, or wrong. Claiming your only is impossible if you're worried about how others will judge you.

QUIET AND UNBOUNDERED

Had Cindy always been this strong? In fact, she didn't always speak her truth. There was one instance in which she kept silent, which is her second story, and one she came to regret.

Early in her advertising career she had to live in Cardiff, Wales—a port city with a big immigrant population—for a work assignment. She had been looking for a cheap place to live and had been having a tough time of it. At one point she was being interviewed for a room by a homeowner "who [saw] me as English" and then began talking about "those Chinks." Cindy thought to herself, *I can say I'm half Chinese, or I can get a room to live.* She kept silent.

Although her options at the time were limited, she remembers the raw feeling of denying herself, her heritage, her truth. That incident taught her what not speaking up meant for her values and her interests. "Today, I would not remain quiet," she acknowledges.

Perhaps it was this experience of silencing herself that encouraged her to be bold at the advertising firm she later joined, which was the setting for the third story of how she discovered her meaning.

One day on the job she found herself alone in a hallway with Nigel

Bogle, the operating head of the firm, and couldn't resist the opportunity to pin him down to ask a question that she'd been mulling over: "Where am I headed in this agency?" To that point, she had had an illustrious global advertising career, having worked at Ted Bates, J. Walter Thompson, and Gold Greenless Trott before joining Bartle Bogle Hegarty in 1980 to run large global accounts such as Coke, Ray-Ban, and IBM.

Nigel smartly turned the question back on her and replied, "Cindy, you tell us what you want to do, and we'll make it happen."

"I wasn't actually prepared to answer the question, only to ask it!" she recalls.

Bogle's parting words stayed with her. "Don't be boundered by the realms of the possible!"

She began to think about what it was she really wanted. Ever since spending a summer in New York City with an Oxford friend in 1985, Cindy had been trying to get back there. It had become her go-to destination for vacations, and she found herself setting up US client visits on a Friday so she could stay the weekend. "The first time I came, I knew I had found my spiritual home. This city has my energy, pace, dynamism. It and I are entirely empathetic."

Cindy used her ideal location as a first design principle. "I came back to Nigel and said my dream job [would be] running BBH North America and working in NYC." At the time, the agency had no footprint in North America, which meant that Cindy would have to establish an office there, a fairly large risk for BBH given that no client was asking for one. Instead, in 1996, when the agency needed a presence in Singapore, they asked Cindy to relocate there. Instead of being disappointed, she saw the move as an opportunity to build something from the ground up, to prove an expansion model could work. She started and ran BBH Asia Pacific, where she ran the Levi's account, and waited the two years before BBH was finally ready to open up in the United States.

In 1998, Cindy found herself alone in a room with only the phone to keep her company, the founder of BBH North America. Four years

later, BBH New York had won clients like Johnnie Walker and Uni-
lever, and was named *Adweek*'s Eastern Agency of the Year. By 2003,
Cindy was voted the magazine's Advertising Woman of the Year and
named chairman of BBH New York.

Her advertising career was successful because of her skill in the job,
but also because she had the courage to say what she thought and to
ask for what she wanted.

COUGARDOM

In 2005, Cindy resigned from BBH, eager to do something different. As
she turned forty-five, she had her "very own personal crisis for what's
next," realizing that, if she wanted to review her options, the best step
to take would be to put herself on the market.

Here I am, world . . . what have you got? she wondered. Given her
reputation in advertising, news of her availability led to some fifty in-
terviews, which were mostly offers to do BBH all over again. During
that time, though, she also founded a start-up called IfWeRanThe-
World, which she conceived of as "an MTV of 2006—uniting young
people around not music, but their desire to do good in the world, to
enable them to translate good intentions into action." Esther Dyson
called it the "liquidity of goodness."

While she was building IWRTW, she was consulting on the side,
which became her entrée into cougardom when she began doing work
for an online dating site and wanted to fully experience the product.
She had been dating younger men for several years and now posted a
profile on the site. As the *New York Times* reported,[7] "To her surprise
and delight, she said, the majority of responses came from younger
men. Ms. Gallop has barely glanced back since."

She conceived of MLNP as the result of something she experienced
and in response to a problem that she saw needed to be solved. Her original plan wasn't to launch a start-up; she was simply going to tell others
what she had learned by sharing her stories and insight about them. "I

date younger men," she would tell audiences, "predominantly men in their twenties. And when I have sex with younger men," she continued, "I encounter very directly and personally the real ramifications of the creeping ubiquity of hardcore pornography in our culture." Here was a grown woman talking openly and honestly about sex. She had taken Nigel Bogle's advice to heart and refused to allow herself to be boundered by the realms of the possible. She gave a name to what she saw, and because she was in a position to do something about it, she acted.

CLARITY OF PURPOSE

Cindy's experiences—of growing up and being told she was less than others, of "passing," of asking for what she wanted at BBH, of noticing young men's sexual behavior—all helped her arrive at what gives her work meaning and serves as the vision for Make Love Not Porn. The thousands of people who have joined her provide compelling testimony that her concept is a forceful dent maker in real-world sex.

But what can her stories teach the rest of us? How can they show us the way to discover one's own path?

"I never consciously planned or set out to do any of what I'm doing now—it was a result of coming across something I felt strongly about," Cindy says. Her clarity, she explains, came "organically, over fifty-six years of living—of noticing, of acting."

Some people conceive of identifying clarity of purpose, of charting one's path, as if it's something to be found, like booty on a treasure map.

But what if the path of clarity is one that is suggested by an observation made by Martha Graham: "There is only one of you in all of time, this expression is unique. And if you block it . . . it will be lost. The world will not have it." Instead of a map, then, which offers directions to well-blazed trails, you need a guide that will help you to navigate a topography of newness. For this, you'll need the skills of navigation, and the tools of orienteering, to head into uncharted territory, where you'll discover your own path.

What are the skills that will enable you to discover your purpose? There are three: (1) Notice. (2) Act big. (3) Act small.

First, notice what it is that you are drawn to, what matters to you. What are the niggling questions that won't let you rest? What is the one thing you really want to do but you're worried someone won't approve of it? What problem do you think needs to be solved? Your focus can be small or big; it can be a specific topic or the way a particular process can be done more efficiently. It can be teaching young people a skill you possess so that they become better at it. It can be inspiring team behavior by coaching the local kids' soccer team.

It's often difficult to be aware of what only you notice. You might think everyone else must have the same concerns, if you do. A good way to identify your own personal "only" is by imagining you were born with a lightbulb on top of your head that shines continually.* It might be a golden orange, or teal blue. When you enter a room, your light illuminates the entire space. That makes it difficult to discern your own only, because the light you shine is also the filter through which you see the world.

Here is where others can help you: When you walk into a room, people can observe, "Wow, it just got orange in here!" and help you name that particular shade of orange. They have the perspective to be able to see the difference in the world when you're present, and when you're not.

This is not to say that you should ask your friends, "What do *you* think I care about?" because that's not their responsibility. What they can do is describe your orange to you, and explain how its radiance affects its surroundings. What that ultimately means to you, and how you choose to act upon it, is your task.

Sometimes aging helps in the noticing, because the lightbulb grows

---

* The lightbulb analogy was created by Justine Musk, who loved the idea of onlyness early on and shared her take on why it matters. "What We Talk About When We Walk About Purpose," JustineMusk.com, May 6, 2013, accessed February 20, 2017, http://justine musk.com/2013/05/06/what-we-talk-about-when-we-talk-about-purpose.

ever brighter with life experiences. That should give hope to those who haven't found purpose earlier in life. Both André and Cindy made some of their most meaningful contributions after they turned fifty. Age is not a limiting factor—and neither is youth.

Second, don't just be you, *do* you. By now, I hope it's clear that there is no one "right" you, one perfect notion of your identity that therefore dictates what you can or cannot do. To say there's a "right" self would be like saying there's a "right" kind of cake, when in reality you wouldn't reject a pound cake for not being an angel food cake. Believing in a single "right" way often causes people to look outside of themselves, to make comparisons to others and find themselves wanting. Charlie Guo, for example, was looking outside of his own history and experiences, visions, and hopes in search of "the right" answer, instead of *his* answer.

There are two ways to conduct oneself in daily life. One is to be competitive/comparative and measure your own value against that of others. In the *contributive* way, in contrast, you focus on doing what *you* can. Comparative judgments and the inevitable anxiety to which they give rise do not inspire you to use your own only. But contributive focus does, because it inspires you to do what you, and perhaps only you, can.

Third, don't give a damn what others think. This advice can be difficult to accept, because at its deepest level it involves addressing fear.

Fear has been a part of every story we've encountered thus far, as fear is a fundamental aspect of the human experience. As Thomas Hobbes expressed it, "I was born and Fear was born with me." Kim Bryant feared for her daughter's spirit, and what the consequences might be if she gave up her job to work full-time on BGC. Zach Wahls, Ryan Andresen, and Pascal and Lucien Tessier feared the stigma and judgment of others. André Delbecq feared rejection for taking on the novel subject of spirituality in business.

They all experienced fear, and yet they were not deterred by it. If fear can stop you, it can confine you to fit in to society as it is instead of shaping the world as you imagine it, into one that has more value for both you and for others. Fear can cause anyone to self-handicap. As Cindy says, "Fear of what other people think is the single most paralyzing dynamic in business and in life. You will never own the future if you care what other people think."

This is not to say that fear doesn't often serve a useful purpose. There are surely things to fear, like snakes or falling from high places. Fear is a signal that can protect you by warning you to be more careful. But your aim should be to distinguish real fears from phantom fears. If you can make that distinction, you can pay attention to genuine fear long enough to appreciate how it can guide you.

Learn to address your own fears, your "darkest and most negative interior voices the way a hostage negotiator speaks to a violent psychopath," wrote Elizabeth Gilbert on this topic, "calmly but firmly. Most of all, never back down. You cannot afford to back down. The life you are negotiating to save, after all, is your own."[8]

I have written about my standing appointments with fear.[9] There are just a few ground rules that we've negotiated over time—I need to be curious, I need to be conversational, and of course, I need to be honest. I need to actually listen and hear what I need to hear. I need to not try to convince fear to go away. Sometimes I get upset, but I try not to defend what I've done or overpromise what I will do in response. By letting fear have its moment, I know it can wait and remain quiet until we have our next meeting. I remain afraid of all the things I fear—but only for a while.

Fear, for its part, has developed manners. It no longer visits at 3:00 a.m. It has stopped consuming me during the thirty or so minutes before I begin a live broadcast with several thousand people watching. It has ceased consuming my energy as I prepare for strategy sessions. It has given up offering side commentary when I'm parenting through tough times. If I ask fear, "What concerns you?" it tells me immediately and usually quite clearly. And it often provides me with an insight that I needed to get better at a particular pursuit. All these things allow me to be more present in the high-stakes situations where clarity matters.

When Cindy says, "Stop caring what anyone else thinks," she includes an idea she recommends to the young people she mentors: "Conduct an experiment for one day: Do everything not giving a damn what anyone else thinks. Say what you really think. Do what you think needs doing. Try it on, like a new garment, for a day or for a week. Bound it [in time] if you're scared."

You'll soon see that discovering yourself is a function of practicing being yourself. But it can't happen without action. "The single biggest pool of untapped resource in the world is human good intentions that never [get] translate[d] into action," as Cindy describes it.

YOU BECOME IT BY DOING IT

How did André or Cindy ultimately discover what mattered to them?

The same way you or anyone would: by examining what they cared about, by relentlessly chasing a question connected to it, and by giving themselves the freedom to envision something new to improve it. But this is only the start. Until you do the actual work, the strength and specificity of your goal will not become clear—to you or to others. Until you do the work, you won't know yourself well enough to signal your mission clearly, which will be a key factor in finding and recruiting your fellow dent makers.

A vignette in J. Ruth Gendler's *The Book of Qualities* compares embracing one's power to trying on something new,* which begins to fit you better the more often you wear it:

> Power made me a coat. For a long time, I kept it in the
> back of my closet . . . I didn't like to wear it much, but I
> always took good care of it. When I first started wearing

---

* This is a book to keep on your desk. From beauty to compassion, from pleasure to terror, from resignation to joy, it is an insightful exploration of the rich diversity of human qualities.

it again, it smelled like mothballs. As I wore it more, it started fitting better, and stopped smelling like mothballs. I was afraid that if I wore it too much someone would want to take it, or else I would accidently leave it in a dressing room.

But, it has my name on the label now, and it doesn't really fit anyone else. When people ask me where I found such a becoming garment, I tell them about the tailor, Power, who knows how to make the coats that you grow into. First, you must find the courage to approach him and ask him to make your coat. Then you must find the patience inside yourself to wear the coat until it fits.

Cindy has worn her life's history, experiences, visions, and hopes—her onlyness—until it fits her like a glove. How do you discover what matters to you? By doing what matters to you. Cindy's story gives us all the confidence to don our own coats, in one big, bold move, and let them warm us from inside.

PART II

Co-Denters

The Power of Meaningful Relationships

By now, you've seen the power of claiming one's "only" as only one can. By doing so, you are able to define not only *what* matters, but *why*. This is your fuel for the journey ahead. With it, you have the capacity to shape your own destiny and to know where you want to go, purposefully. Now you can do something—whether it involves taking a small step to address a local issue or starting an entire global movement—to contribute the best part of yourself to the world.

This is the first step to being powerful.

The next is to find potential allies with whom you can join forces. You no longer need to have a position of authority within an organization to get people to follow you. But you do need the strength of a community enlisted in a common cause. It's important to be deliberate in your choice of supporters, selecting those who will challenge you to be your best while standing with you as you try (and sometimes fail) to do so.

At first glance, this might sound like a simple matter of building relationships. But is this ever a simple process? Sometimes relationships are shallow, based on things that scarcely matter, and effectively leave you alone with your convictions. At other times, they are unaligned, with each party having a different goal, which prevents momentum from building around an idea. They can also be unsupportive, which limits the risks you'll take and the creativity they'll inspire.

Identifying others who are "like you" means more than just assembling networks, making contacts or friends—it means finding genuine collaborators. Knowing how to find "your people" takes a set of particular skills. To find them, you have to know how to both *signal* your passions and interests and to *seek* out theirs. What are you actually looking for in others? Where will you be likely to find them? How will you reach out to them (in person, or via e-mail, LinkedIn, Google, or Twitter)? Chapter 4 tells the stories of Alex Hillman and his fellow Philadelphians, who learned what it meant to be more themselves as they navigated different types of communities to find "their people," and of Rachel Sklar and Glynnis MacNicol, who sought and signaled those who shared their deepest purpose to create a tribe of the like-minded.

Chapter 5 considers how to reach agreements on shared purpose, even if "the other side" doesn't seem very convincible. Franklin Leonard challenged an entire industry to reimagine what it was, merely by asking a new question. Meanwhile, Samar Minallah Khan, a feminist filmmaker, created an unusual alignment of purpose with Pakistani tribal leaders to stop little girls from being traded as compensation for crimes.

Chapter 6 explores the most complex issues of forming an "us." Relationships are to the modern Social Era what efficiency was to the Industrial Era—the foundation for the kind of scale that makes it possible for individuals to work together to accomplish great things. Relationships are fundamentally based on trust, which gives us assurance that we can lean on one another and unite not just in principle but also in practice. Tom Rielly shows us how to create the context of trust, showing that belonging comes at a personal cost, an investment of self. While trust has eroded in people and in institutions, we all need to understand the way we come to count on one another. In the story of PatientsLikeMe, we see the ways the fundamental trust equation must be in balance for this lever of *us* to work.

CHAPTER 4

Find Your People

*When we are really ourselves—when we really connect with
who we are and what we care about, and we have the
confidence and the support to be forthright
and honest—we find each other.*

—PETER SENGE

INDY HALL: CREATING THE CLUBHOUSE

When students at the University of Pennsylvania were asked in 1990 if they would consider a career in the city of Philadelphia, where the school is located, only 35 percent of them said yes. By 2010, that number had risen to 60 percent.*

Some of that increase is due to Alex Hillman and his clubhouse for weirdos like him. That project began because Alex was lonely. Although he was surrounded by people, he wanted to be around others

---

* Nationally, about 42 percent of students stay in the area where they go to college; in the Philadelphia area, 64 percent now stay: Deborah Diamond, letter to the editor, *Philadelphia Inquirer*, August 23, 2016, http://www.philly.com/philly/opinion/20160823_Letters __Making_Philly_a_magnet_for_millennials.html.

who were "like him," but, not encountering them anywhere in Philadelphia, he figured they must not exist.

Alex had been thinking of leaving Philadelphia for the West Coast, where all the other creative tech types seemed to be heading, and had a connection to a tech company for a full-time job. The California-based business represented all that he wanted: creative work, smart peers who loved to take on challenges, and an opportunity to do something that would have an impact.

"I was a rarity in Philadelphia, or so it seemed. I felt like I was hitting a ceiling of advancement where I worked, and mostly I just felt alone in my field, in my work, in my city," Alex explained. Before he dropped out, he had spent his college years living in a bubble, staying mostly on campus. His first job, as a Web software developer, was in a suburb an hour outside the city. There were some local communities, of course, but when he checked out the existing organizations and players, none of them felt exactly right.

"What I had found was old school, parochial, and led by people or institutions who were more interested in preserving yesterday than creating tomorrow. Lots of money and power seemed to come from very few sources," as Alex recalls. Some of these organizations had grand yet vague missions of driving technology growth in the city, but when he went to meetings for those organizations, they "couldn't point to even one example" of something new or interesting they were actually doing.

When the company Alex was interviewing for in California "couldn't quite get their shit together," he began to reevaluate and decided that he ought to give himself "one more chance to find a way to stay in Philly," which still appealed to him.

Giving himself a six-month block of time, he set out in January 2006 to find and meet as many like-minded people as he could for potential colleagues. "A big part of me thought that this would be fruitless, that what I was looking for didn't exist," Alex explained, "but it occurred to me I was looking in the wrong place, or I was looking in the wrong way."

While Meetup, Scott Heiferman's platform for self-organized

groups, had existed since 2002, it had little presence in Philadelphia. Alex augmented it with his own legwork—researching professional associations, finding small conferences, attending tech-based events in the city—to find other "geeks like him."

"Whatever [potential] watering hole there was for animals like me, I went."

As he went from event to event, he started spotting the same people and found a few who seemed to have a similar outward profile— tattoos, ironic T-shirts, checkered shirts instead of suits—to himself. As he got into conversations, he discovered that he had a great deal in common with some of them.

These creatively geeky types were all equally relieved and happy to find one another. "There were quite a few of us looking for the same thing," David Dylan Thomas, who directed and coproduced the Web series *Developing Philly*, recalls. "We all wanted the value of working with a great company like Google without working for a big company like Google . . . and with this gathering, we were getting the quality of peers, without having to join up to a big firm or leave our city."

In September 2006, this crew of coders, makers, and designers began meeting every Friday in cafes and coffee shops, where they remained for the entire day and soon discovered that they were actually more productive as a group. "We could just look up and ask for help on a technical glitch, or ask if anyone else knew how to do X, Y, or Z. Our collective skills were strong," explains David. The more time they spent together, the more they learned from one another, were inspired by one another, and pushed one another. After a while, they began gathering on other days of the week as well.

Communities like this cannot be forced coalitions—you can't simply decide, "We will be this to one another," which would negate onlyness, a connection made by choice.

The most rewarding and productive groups come to *be* something together, and for each of their members, as they come to know one another.

FINDING A SHARED PURPOSE

"We had built a club, so we needed a clubhouse." Alex pitched this idea during a five-minute ignite-styled* talk at Creative Camp, an event sponsored by a local recruiting company. A few dozen people attended, many of whom were part of Alex's posse, so he decided to give voice to something that had been brewing at the back of his mind. He shared what he knew about coworking and said it was time for Philly to have something like a clubhouse where the creative geeks, solopreneurs, and entrepreneurs in Philly could gather.

The group quickly embraced Alex's idea and added their own suggestions. Near the end of the session, a user experience designer named Lauren Galanter spoke up: "I love how you're not talking about it as simply a place to work, but a better version of working in Philadelphia. If you end up opening a coworking space, you could name it something with a Philly theme, like Independents' Hall. You know, like the historic site, but a hall of people who are independent." While Alex was the catalyst, it was ultimately the community that created Indy Hall, the first coworking space in the city. Long before they occupied a physical space together, they co-owned the idea as their own. They all occupied a certain geek*ness*. They occupied a particular creative-*ness*. They occupied doer*ness*, and occupied Philadelphia*ness*.

In their shared "nesses" they coalesced as a community united by common purpose.

It was the group that painted the walls, that assembled the Ikea desks, that chose or made art. While Alex was a cofounder and was responsible for writing checks, Indy Hall was cocreated by many. Indy Hall was eventually recognized by *Business Insider* as "one of the coolest

---

* Ignite talks are a way to hear ideas quickly. The format is standardized: five-minute-long talks during which each presenter must use twenty slides, which auto-advance every fifteen seconds.

co-working spaces in America, a hotbed of intellectual activity, seren-dipity and expansion of one's networks."[1]

SIGNALS IN THE DARK

How is it that Alex went from feeling like the "only one," thinking of leaving Philadelphia, to joining with nearly twenty people to create a communal working space? Had he simply been oblivious to the possibil-ity earlier? Had none of the members been interested in such a project beforehand? Was it that he simply needed to look harder for like-minded people?

The answer lies in the crevices between these various questions.

Parker Palmer, the philosopher and author of *Let Your Life Speak*, wrote, "Long before community assumes external shape and form, it must exist within you. Only as we are in communion within ourselves can we be in community with others."

This tension is important. Onlyness—the word itself—conjures up for some the idea of a singular hero. But the underlying power of it follows the duality inherent in the word "individual," which is the smallest member of a group. An individual is therefore never isolated; he is always con-nected. Onlyness contains a similar duality: It is born of you *and* unites you meaningfully with others; it is the connected you. When Western society romanticizes the image of the individual, it denies the power of *us*. This distorts the process of how any of us actually gets things done, and the understanding of how much we need to belong to one another. This distortion allows those in power to maintain the status quo, because it keeps those who have wild ideas on the fringes, separated and alone.

The account of Alex's search for "his people" offers a few other spe-cific suggestions for how to signal your interests and assemble your own community.

Alex began by personally claiming something as mattering to him. He signaled his intention to find people who shared his professional concerns. Then he acted on those concerns as he sought out venues where passionate technical people might be already gathering. It's not coincidental that, as he stopped wearing button-down shirts when he really wanted to wear a sweatshirt, he showed up as himself.

What happened next demonstrated the power of focused *signaling*, which is to convey information about a specific area. Indicating his interests clearly encouraged others to *surface*—to rise up and signal their own interest. Many creatives wanted to find one another; they shared the objective that seemed deeply personal to Alex. Signaling and surfacing thus brought a disparate group of individuals together into an organized whole. It is as if you look at the night sky to see many stars, but no pattern. And then, someone comes by to show you the orderly arrangement of the Little Dipper. Once it is visible to you, it's visible. But until the distinctions are surfaced, all the stars seem random and undifferentiated. Signaling and seeking between people is the invisible cord of meaning lassoing people together into an organized whole.

The Invisible Cord of Meaning Connects

How you actually do the signaling matters. Alex invited people in Philadelphia's Creative Camp to create a better type of communal gathering place. He did so not by making the idea only about him but by presenting the idea in such a way that others could stake their personal claim to it. The former is a tight fist wrapped around an idea, the latter an open palm holding the idea so it enables others to pick it up and hold it as if it were their own. This stance allows for people to co-own the idea, which is central to onlyness.

The term "follower" is often used to designate the members of a community, as if it is an entirely passive role.* But a more specific description would be "joiner." To join is to be one among equals who want to make something together. Alex's "leadership role" here was to ask a new question and create the space for the rest of the conversation (and later action) to ensue. The answer was arrived at by many, together.

Keep in mind also that signaling and surfacing is an organic process, and one that does not advance quickly or by outside pressures. There's always a temptation to give up too soon in the effort to find others who care about the same things, or to push too hard at the beginning of the quest instead of paying careful attention to what draws people toward the common purpose.

FIVE FORMS OF COMMUNITY

Going to meet-ups is one way to find people like you. But you can also search Google, or scan LinkedIn profiles and send "Connect with Me" notices. You can tweet links about what you're interested in or leave comments on an Instagram feed.

The particular qualities you value that will make someone a potential

---

* Underlying most models of followership is some sort of dominance and some sort of deference. I adhere to Barbara Kellerman's approach to followership, which uses level of engagement as a way to show followership can be passive or active. A good resource to understand followership in organizational contexts is her article "What Every Leader Needs to Know About Followers," *Harvard Business Review*, December 2007, https://hbr .org/2007/12/what-every-leader-needs-to-know-about-followers.

"like you" will affect how you search, what you search for, and even where you search. What follows is a taxonomy that helps in that effort by identifying five different types of communities,* based respectively on practice, proximity, passion, providence, and purpose. To be sure, a community and a logically grouped collection of people aren't the same. Until you have shared purpose and the formation of trust, you won't have an onlyness-based community. But seeking out groups is the way to start finding one.

Practice: This form of group is united by an activity in which they all take part. Alex provides a perfect example, in that his colleagues were all creative types who were interested in engaging in Web 2.0 development. Other examples are entrepreneurs, Web designers, filmmakers, venture capitalists, and librarians. Practice is not limited to paid professional activities; it applies as well to anyone who wants to take part in a common interest or hobby, like French speakers or marathon runners. The most efficient way to find others who have such mutual interests is through an *online keyword search*.

Proximity: A group based on being *of* or *in* a certain place. During my recent relocation to Paris, I learned of a website called Message Paris. Members paid a nominal subscription fee to find others in the city and get quick and easy access to such information as how to choose a school or how to deal with foreign taxes. Later, I found another website called Mes Bonnes Copines ("my good girlfriends"), which connects people to one another to offer various kinds of helpful services, such as swapping babysitting services. I also used geolocation on Twitter, focused on Paris and even on my particular arrondissement, which led me to information about upcoming events, local news, and people

---

* Harry Max, one of the most brilliant coconspirators that I've ever had the chance to work with, first came up with this formulation of the Five Ps construct while he worked for me at Rubicon Consulting, a strategy firm. If you're interested in the research on communities, there's an archived copy of the original paper on my website: Harry Max and Michael Mace, *Online Communities and Their Impact on Business: Ignore at Your Peril* (Rubicon Consulting, October 22, 2008), http://nilofermerchant.com/wp-content/uploads/2008/10/Rubicon-web-community.pdf.

I might find interesting, for example, who won the "best boulangerie" contest near me. These resources helped me to navigate a new city and make contact with others who were doing the same, using the parameters of *territory and geo searches.*

Passion: A community of passion is driven by a shared interest in a particular subject, but differs slightly from one based on practice. Let's say you're a parrot lover, as my friend Arikia Millikan is. Her journal— she's a writer, so her notebook will likely be in plain sight—has loads of parrot-related content, so whenever friends see it, they can't help but notice and say, "Wow, you must *really* like parrots." Even though most of them have never seen her actually interact with a parrot, they send her all the parrot videos and paraphernalia they find. Arikia also shares anti–animal abuse petitions on her social media pages, so people come to understand that the topic matters to her. A combination of digital tools has helped link her to others "like her" who share this passion. This same method can be used for dog lovers, motorcycle enthusiasts, pottery makers, art collectors, skiers, and so on. Searching based on passion is *content driven.*

Providence: Provident communities are the product of seemingly random connections—the serendipity of meeting just that right person whom you later went on to found a company with, or high school buddies who introduce you to your future funders on Kickstarter. This process is actually not as random as it appears. For example, entrepreneur advocate Tereza Nemessanyi was not a known expert when she first dived into that world. She started by first soaking up ideas from blogs written by venture capitalists but then noticed that lots of interesting ideas were actually published in comments sections. She began to leave her own comments, which led her to greater personal clarity and led others to discover her. She then created a blog on her own domain, tweeting actively to support it, which after three years led to her building her own company. As she attended conferences promoting her business, she found people already knew of her from her comments and so were interested in talking with her. After she left her own start-up, she got a job at Microsoft, helping the company to build strategic relationships with start-ups and entrepreneurs. Providence is, by definition, random. But there is clearly a strategy to leverage it, which is to *figure out where you can be* to create opportunities for serendipity, like conferences or blogs where the topic or attendee self-selection process creates a fruitful context.

Purpose: Purposeful communities are those that share a vision of the world as it could be. Identifying people based on their interest in a specific goal or challenge is the hardest way to align with a community because they are not typically found in one spot, or by keyword searches. Even those who appear to be alike enough to have shared purpose won't always have that alignment. Unlike practice areas, purpose-oriented communities don't always have known user groups and are, more often than not, not geography dependent. The story that follows later in this chapter about The Li.st digs in to what is distinctly needed to build a community of purpose so you gain the strength of many without dilution of meaning. Because these groups are issues based, finding people with common purpose means finding those who *share a commitment to a cause*, and that involves a more sophisticated method of signaling and seeking.

This taxonomy of communities is a framework you can use to think through whom you need to find and how best to reach them. If you were, for example, looking for fellow parents trying to change an after-school program, you could find them based on the practice of child-focused after-school programs and also based on proximity. If you were looking for people who were interested in biosciences and entrepreneurship, you'd look for a mix of practice areas so you could find the subset of the groups that met this combined criteria. It's not meant to be definitive as much as a place to begin, so feel free to mix and match strategies that work based on your particular needs.

A PLACE TO BE, NOT A PLACE TO BE FROM

A tattoo on Alex Hillman's right forearm reads "JFDI," which stands for "Just F-ing Do It." It's Alex's own signal to himself to act, to do, and to get over fear.

Today, Indy Hall is the working home of more than three hundred people, all of them committed to both "having a place for people like me" and supporting the local tech scene. It's more than just a collection of desks, because its members share a set of values, beliefs, and dreams. The space is a vital thread in the social fabric of the start-up culture, and also of the city. Philadelphia is now a place to *be*, not a place to be *from*.

The city's former deputy mayor, Rich Negrin, confirms that Indy Hall is a vital part of the community. "Citizenship is a full-contact sport," he explains. "Cities are fundamentally about relationships. Cities only work when we serve our shared goals, share our thinking, [and] share ourselves. It's not what I do *to* or even *for* you, but what *we do together*. And only when we have vibrant community organizers can we have a vibrant city. Alex and Indy Hall are just that. They help one part of our community to build relationships with one another and, ultimately, to the city."

Philadelphia now has a sense of place. The blight portrayed in the *Rocky* movies has been supplanted by vibrancy. As of this writing, Philadelphia has had six straight years of growth. Code for America is sponsoring its second major deployment there (Philadelphia is the only city that has hosted two events).

On Alex's right forearm is another tattoo. That one is a musical symbol, a fermata, which signifies a pause. It is a space that allows performers to reconnect and remain in synch with one another, so they can play and perform together.

Just as Alex has done for the community.

SHOUTING IN THE PUBLIC SQUARE

People seek and signal all the time, but sometimes in ways that discourage rather than encourage connection. You won't get anywhere by spamming strangers. You won't find your co-denters if you stalk them—which brings us to the topic of what *not* to do in the creation of a community.*

When someone you've never worked with or even met sends you a "Hi, I'd like to add you to my professional network on LinkedIn" request, do you accept it or let it sit? When someone doesn't write back after you post a Facebook message, do you simply hope they'll get to it, or do you send another one the following day? If you're raising funds for a new project on Kickstarter, do you simply hope that people who might find it interesting learn about it, or do you urge twenty people via their Twitter feeds, "Look here!"

We all want to connect, communicate, and, ultimately, build stronger relationships that help us get real work done. We want to find people to run with. Social media tools hold that promise, but too often online behavior becomes self-promoting, loud, and unhelpful. That's when people run away from us. So, what protocols and approaches help us be more connected rather than less?

While most of us want to come across online as someone worthy of friending or liking, and certainly of paying attention to, at times we're all prone to appearing awkward, if not goofy. If there's an analogy to draw, it's that the social world online is like a new school, and most of us are acting very much like adolescents in our first year there, trying to figure out the unwritten rules. Most of us inevitably succumb to some form of "what not to do" that gets in the way of effectively making human connections. For example:

---

* You've probably already noticed each chapter has a "what not to do" story. This concept was inspired by Stacy London's television show *What Not to Wear*. It's not to make fun but to see what it looks like, so we can do better.

USING GRAY, FACELESS ICONS. What does your online biography look like? We need to see you—not a gray, faceless icon, but *you*. If you have an egg as your Twitter avatar, a shadow picture on LinkedIn, or some Second Life–style digital rendering for a profile picture on Facebook, it suggests you're hiding. Even using your cute dog or cat to represent you is a form of masking yourself. Connections are made by seeing actual human faces, a response that's wired into our human brains.

LIKING EVERYTHING. The person who favorites or likes everything is simply transmitting a lot of noise rather than a clear signal. The message this habit often conveys is that you have nothing else to do, or very low standards, or that you're a sycophant. In any case, more than likely it means you lack authentic opinions. To like everything is to be a pig in a trough. If you indicate what you genuinely like, people will know who you really are. Your onlyness will shine through and connect you to what matters.

MAKING IT ALL ABOUT YOU. It's okay to promote yourself—once a week, or even better, once a month. After you do so, move on and pay it forward by promoting other people and their ideas. Constant self-promotion is a turnoff, as is posting a lot of material every day, which makes other people's feeds all about you. Turn your own feed from a channel about yourself into a showcase for ideas that you think matter. Share something that many of your followers might care about, and tell them why you're doing so. That's useful signaling of your interests.

TURNING TOOLS INTO PITCHES. Hashtags work to index information on Twitter so that people can search for similar content. For example, #OSFest16 was the hashtag for a conference that assembled a thousand collaborative-economy thinkers. If you were one of the attendees of that gathering, you could use it to find others who were present. If you weren't attending but wanted to know what key topics were discussed, the hashtag could direct you to them. Using multiple hashtags, however, indicates you're likely trolling for new users, or peddling your wares.

PLEADING FOR ATTENTION. There ought to be a word for those who send incessant requests to others asking them to promote or retweet or amplify their work. Anil Dash, who has half a million followers on Twitter, shared that people often tweet to him: "I wrote a thing!" or "Your followers will love this!" or "Can you just share this real quick?" He explains that "the volume of these requests is typically in inverse relation to their merit. Sure, some of these are cool things I'm thrilled to get to share with a (theoretically) larger audience. But the overwhelming majority is just crap, or things that nobody would believe I was sincerely sharing. Worst of all: Nobody clicks."[2] In my own in-box, a professional friend regularly shared things with me, asking for a tweet. After five such requests, I got annoyed; after ten, I was resentful. By the eleventh, I stopped reading anything from her; her e-mails now get moved to spam. She not only lost my attention, she lost an ally.

INVESTING NEVER, REAPING NOW. Today, social networks are incredibly important tools. They serve as our virtual watercoolers, town halls, and community centers all mixed in one. We all know people who insist that they'll never use them because they "have real jobs" or "have a full life." When they do need a job, however, they suddenly discover the power of social media, having invested nothing in it. That tactic never works. If you're the kind of person who never needs anything—a job, knowledge of your industry's latest developments, advice on how to raise your baby, or a guide to the best local restaurants—then feel free to skip the whole social network scene. But at some point you will be likely to need help from someone knowledgeable, so participate. You can do so in small ways, for a minimal amount of time, but don't opt out completely.

BEING STALKERISH. If you contact someone and reveal that you have lots of information about him, he's likely to be suspicious. Remove that uncertainty with clarity about your intentions: "Zoominfo says you used to work where I worked and our time there overlapped." Researching someone's background is fine; just be upfront about it and why you're doing it.

THINKING HOMEWORK IS FOR OTHER PEOPLE. At the same time, if you don't make the effort to devote at least ten minutes to establishing a meaningful connection, you show you don't care. When I wanted a friend's particular insights on this book, I spent time checking out her recent Facebook and Twitter activity and discovered she had some exciting personal news, which not only provided a good entry point for a conversation but indicated my respect for her.

MISUSING CHAT. Use chat as if you were standing next to the person you're conversing with: Think lunch plans, brief gossip, and quick yes-or-no questions. Don't inundate anyone, and please don't send attachments or YouTube links in chat. Chat (and texts) are for immediate communication; if you want someone to remember something beyond this tiny little sliver of a moment, then send it to them in a more permanent way, which, for most people, means e-mail. Keep in mind what each medium is best suited for and use it accordingly, or ask the person you're communicating with her preferences.

BEING AFRAID. Most of us learn pretty early on that it's not always safe to be who we truly are. With teachers, parents, bosses, and classmates, we find ourselves editing what we share, fearing that we're not good enough or that we'll be marginalized, or ignored, or even disliked. (There's plenty of evidence that this is more than just a concern for some groups of people—it's a truth.) Those who are part of extended communities may worry that sharing themselves can lead to too many people wanting something from them. These are all just different versions of the same fear: not knowing *how* to be real, to be genuine, and to be our true selves. Online communication only magnifies that core issue. Acceptance and belonging are what we all desire, but to find them, you have to let people actually see you—all of you, in all of your onlyness.

As Sherry Turkle, the accomplished MIT sociologist, has written in her book *Alone Together*,[3] we are shaped by the mediums in which we exist. While we've recently experienced a shift from predominantly face-to-face to link-to-link interactions, this does not mean the breakdown in human civilization that some have suggested. Our ability to seek and signal to a community means, in its simplest form, that you can now be intentional about how to use the tools that create digital ties. The online world is our modern public square, and it's there that people from all over the world signal their interests and the level at which they care about those interests. While not everyone participates online, or on social media, those who do are often the very individuals who want to make a dent in the world. Research from the prominent Pew Research Center indicates that only 14 percent of the Twitter community actively posts.* Contributing makes you one of the ones shaping the conversation, so join in, and do so wisely.†

---

* Research in late 2012 suggested that only 23 percent of the population (rather than the more commonly understood 90 percent) are fully passive "lurkers" of content, while 17 percent of the population could be classified as intense contributors of content. See Holly Goodier, "BBC Online Briefing Spring 2012: The Participation Choice," BBC.co.uk, May 4, 2012, http://www.bbc.co.uk/blogs/bbcinternet/2012/05/bbc_online_briefing_spring _201_1.html.

† Some groups participate at a higher rate on the Internet. In Jen Schradie's 2014 doctoral thesis, she shares research that, in 2012, 94 percent of college-educated Americans used

Remember, though, that searching online is simply one way to find your people. Alex did much of his searching with actual legwork, using online content to supplement it. Through trial and error you'll find the right mix for your dent, but whatever method you choose, be sure to remain connected to the fabric of humanity by your own thread of onlyness. Tug on it to let others know what interests you and what you care about.

the Internet, but only 43 percent of the people without a high school education were online. Given that content creation is labor intensive and unpaid, it is likely to be done by certain economic classes, skewing who is participating in key conversations. See Jen Schradie, "This Is (Not) What Democracy Looks Like: The Internet and Democratic Participation Practices Among Labor and Social Movement Organizations from Different Political Orientations and Social Classes" (PhD dissertation, UC Berkeley, 2014).

THELI.ST: MAKING WOMEN VISIBLE

Rachel Sklar was seated in the back row of a famed conference, watching speaker after speaker and thinking, *Why are they all dudes?* She turned on her mobile device and shared this private fuming publicly via Twitter, adding a hashtag for the conference itself. Within minutes, people following the conference, strangers to Rachel, tweeted back, signaling that she was not alone in her feelings. Among the two thousand people at the conference itself, about ten shared Rachel's frustration. After finding one another online, they later convened in person.

As well as citing the name of the conference, Rachel's tweet included a second hashtag: #ChangetheRatio. That signaled the dent she was seeking to make: to create equality for women in business, whether in boardrooms, on stages, or in the entrepreneurial sphere.

But what is the foundation for a strong enough network to make a dent? Unlike Alex Hillman, who was drawing on communities of practice and proximity, Rachel was trying to create a community of purpose, one committed to making a social change. As such, it could easily draw the ire of those in power. Her story is a lesson not just in how to build a viable network but also in how to deal with the deflectors who will undoubtedly challenge your very motivation and purpose.

FIVE AND A HALF WOMEN

#ChangetheRatio was first inspired by a magazine article.

In April 2010, *New York* magazine ran an extended profile piece on the emergent local tech scene,[4] describing how the city's tech players were gathering momentum and "changing the world." Accompanying the article was a playfully shot two-page spread of fifty-three of the leading figures in this movement.

The photograph troubled Rachel. She had read the piece with a gathering sense of frustration because "while it talked to the lack of women in tech, it only perpetuated the situation; it didn't change it.

"I counted the women mentioned. And, then, counted the women in the picture. Only six women were pictured. In fact, actually only five and a half were, because in one of the shots, the founder's foot was such that it partially obscured the face of one of the few women, that of Teresa Wall, who was on the founding team of Meetup."

These statistics didn't match Rachel's own on the ground experience. A Canadian who had lived in New York City for many years, Rachel was a founding editor of Mediaite (and before that, a writer for the Huffington Post), which put her in the hub of "covering the New York tech scene," so that she "easily could have named another fifty women to be a part of [the *New York*] piece." It also irked her that some male-led companies that were only at the PowerPoint presentation phase were listed, when viable, revenue-delivering software or hardware companies led by women were not.

"Women in the industry don't get their due or make the list," Rachel argues with her rapid-fire cadence. "Without being seen, you can't get the meetings, which can't lead to funding, and so on. Visibility begets access, which begets opportunity, so this persistent gender inequality in the media perpetuates a problem.

"Just because they don't see the deep pool of women entrepreneurs doesn't mean they don't exist. If anything, what this kind of reporting and conference approach does is render invisible those people who are already there." A lawyer by training, Rachel makes her case: "By not being listed, they are effectively inconsequential, erased. Unseen. I've done enough media coverage to know that the next-day story would be the typical 'Where are all the women?' articles," which, as she points out, only reinforces the status quo.

"I couldn't muster the energy for outrage, once again. I just wanted a solution." Her solution wound up involving many others who, like her, shared this goal.

Starting with her near-constant e-mail companion, Glynnis.

IN THE LOWER LEFT-HAND SCREEN SPACE

Change the Ratio, as the project was first called, would eventually grow to have a membership of five hundred, but it began as a network of two.

Rachel first met Glynnis MacNicol in a writing class in 2002, when Rachel was still working as a lawyer and Glynnis as a waitress. "When Gmail came about around 2005, and Google Chat was introduced, Rachel was the first person who showed up in my lower left-hand screen," recalls Glynnis. Where Rachel magnetically draws in crowds, Glynnis is the thoughtful organizer. Where Rachel is effervescent, Glynnis is quietly strong. "Like everyone else whose lives exist primarily online, Rachel and I had a running conversation on Gchat," Glynnis, now a professional writer, explains. "Because we were both covering the media, and to some extent politics, much of that conversation involved us going back and forth on what we were seeing and reading, and what we weren't but thought we should be." Later, they each reported on the 2008 election, Rachel for the Huffington Post and Glynnis for Mediabistro and *Playboy*. At a time when blogging influence was reaching critical mass, the two women were making a significant impact, and their shared interests drew them even closer.

After fuming about the *New York* piece, Rachel naturally turned to Glynnis, who confirmed her instinct that it was time to do something about it. We all need such reliable sounding boards with whom to share ideas, to develop early concepts, and to test them out as they take shape. That dynamic has to be a safe one, because the wrong type of confidant can judge early and yet evolving ideas as being a little too weird, or simply ill considered.

Without a trusted confidant to turn to, we sometimes can't clarify our own thoughts, as few of us are initially as clear as we need to be in our ideas. Heck, without that first person to witness your passion, you might think it doesn't exist.

In Derek Sivers's now-famous TED talk[5] on how to start a movement, he discusses a video where, at first, a single lonely person dances alone in the grassy area of a concert stadium. When a second dancer joins in, the pair adjusts their moves to get into a common groove. A third person soon participates, and then a flood of others who had previously been only spectators. The first follower effectively transformed the lone dancer into a leader, allowing the others who had been watching on the sidelines to feel safe enough to join in.

Glynnis made Rachel's idea not only stronger, but safer.

MEET ME

"Let's get in front of this; let's change the ratio." That was how Rachel invited nineteen other women she knew from across the media landscape to join her dent.

"Things aren't changing organically, and if we don't do something, we'll be having this same conversation five, ten, even twenty years from now," Rachel told the crowd when they assembled at a bar in New York's East Village. Not everyone on the e-mail chain could come in person that night, so the conversation continued on e-mail, with people agreeing that it was time to move beyond a single venting session. "And then we had another meeting at someone's house, and then people were added onto the e-mail chain. It grew and grew."

Small groups are a natural extension of the trusted individual as a vehicle with which to test out an idea. Like that first coconspirator, small clusters provide a safe space in which to explore an idea, to hear yourself think out loud, to work out the kinks of your own argument, and to feel supported enough to take the risk to own your own "wild" idea. Because the members of these groups already know you're competent or have a similar perspective, you don't have to worry about being judged for concepts that aren't quite fully formed.

To scale an idea and make a significant dent, though, you have to expand beyond the close-knit group to build a more extended network, and then to add shared activity.

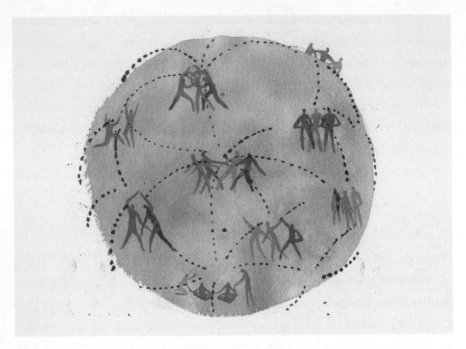

Join Extended Networks

That means not just involving *more* people but connecting different circles of people who share a common cause, not just commonalities. To create scale, you'll need to bring in the onlyness of many, but many who are "like you" in a meaningful way. The challenge is doing so without undermining the safety of the original group.

Just as the conference venue created a context for Rachel to engage a topic, any new event—a news story, the publication of a new book, a blog post—can provide a similar opportunity for signaling and seeking allies. In Rachel's case, another such occasion presented itself when Penelope Trunk, an *Inc.* columnist, wrote an essay on TechCrunch in October of 2010 titled "Women Don't Want to Run Startups Because They'd Rather Have Children." Trunk argued that women as a group were not cut out to be CEOs.

The resulting online uproar amongst a small set of business leaders, entrepreneurs, and women was vehement. Most contended that one anecdotal experience did not data make. Others took issue with

Trunk's wholesale dismissal of the concept of feminism. I wrote an essay on my own blog titled "Fem-nomics? Or Leadership?" I had recently shut down my own start-up, where I served as CEO, because of some personal family issues, and made the point that Trunk's argument was less about gender and more about lack of leadership skills. That piece caught the eye of Rachel, who had been observing the online conversations around Trunk's piece with great interest. A day after I published my piece, Sklar reached out to me and invited me to join her project. While I hadn't considered what I was doing in writing my blog post as signaling, it served exactly that function. Rachel's initial e-mail started us on a multiyear conversation that connected us in common purpose.

Which speaks to a truth: You signal who you are all the time. You do so in person and, of course, you do so online. When you signal clearly, for example, by writing publicly, you certainly take a risk, as you might offend someone or "out" yourself in a way that causes others to abandon you. But clear signals also enable you to make that key connection to those most like you. If I had, for example, sat quietly on the couch at home, never sharing my perspective on why I thought Trunk's argument was flawed, then Rachel and I might never have met. Remaining silent means you never have the opportunity to join and support something you believe in.

The value that Rachel and Glynnis are adding is one that has been missing in the ecosystem they're addressing: an intradisciplinary network of people who want to see this problem of the representation of women in the workplace solved. By building this network, they are increasing the information flow, as well as coalescing a diversity of skill sets, and therefore putting power behind creating solutions. They are closing what in network theory is called a "structural hole," a gap in information.* When a structural gap is filled, not only does information flow freely but the goals of the group advance as well.

---

* Structural holes are a concept from social network research, originally developed by Ronald Stuart Burt. A structural hole is understood as a gap between individuals who

JOINED TOGETHER

Rachel, meanwhile, continued to enlist allies, growing the original #ChangetheRatio group from twenty of her friends to a hundred committed people, vetting each one individually. Not all of the initial members remained; some left because they had originally joined to support Rachel personally, so the more purposeful the organization became, the less they felt connected to it.

Today, there are close to five hundred dues-paying members of what is now called TheLi.st. The addition of members with different perspectives and from diverse socioeconomic backgrounds, experiences, and locales has been incredibly enriching. When one of the TheLi.st participants made an argument that linked feminism to being a mother, another member, Natalia Oberti Noguera, the founder and CEO of Pipeline Angels, stepped forward and reminded everyone that not all women were mothers, and urged that the tent of feminism be lifted to include all women. Natalia, who is a cis queer Latina, regularly challenges the group to take into account issues of color, gender, and sexuality. Her particular outlook brings the group something powerful, as it expands its core concept to be more complete and to ultimately reach places it needs to reach to achieve its greatest potential.

Ellen Chisa, a TheLi.st member, credits it for expanding her take on her standing within her profession. "I used to think that if you weren't in engineering, or product management, you couldn't call yourself a woman in tech. But then I joined TheLi.st and my idea of who we are grew; it became clearer how much women in various fields across tech have in common."

The more people who join TheLi.st, the greater the amount of in-

---

have complementary sources of information; by closing the gap, they enhance their access to power. A very readable resource on the subject is Brian Uzzi and Shannon Dunlap's article "How to Build Your Network" (*Harvard Business Review*, December 2005, https://hbr.org/2005/12/how-to-build-your-network). J. Kelly Hoey adds her take on the network era in her book, *Build Your Dream Network: Forging Powerful Relationships in a Hyper-Connected World* (New York: TarcherPerigee, 2017).

formation sharing, on topics ranging from how to negotiate a raise to how to get a proper 409a valuation to how to structure start-up employee compensation to how to deal with challenging business partner situations. Some questions are those that can't be easily raised in public: One start-up tech entrepreneur related how a particular venture capitalist smelled her hair at a meeting and asked, "Should I look past it?" She found out this particular VC smells everyone's hair, men's and women's. Projects like TED talks, or profiles in the *New York Times*, are organized. Five women became *Harvard Business Review* columnists when a door was opened for them there. One entrepreneur raised a great deal of money after sharing with TheLi.st that a potential investor had a very unethical side.

Membership in the group also relieves the sense of isolation. "Women often feel so professionally lonely. This is probably the case with any group that has been marginalized," Glynnis says. "When you're not the majority, and the narratives around us don't include us, it's easy to think you're the only one who feels the way you do. But after you see four hundred examples and as many varied stories, you feel less alone."

Strength comes not from the specific numbers, but the sense of solidarity.

TheLi.st isn't the only group dedicated to its cause. Aminatou Sow, who in 2014 was named one of Forbes.com's 30 Under 30 in the tech sector along with her business partner, Erie Meyer, founded Tech LadyMafia in 2011, shortly after TheLi.st began. As *Elle* magazine wrote, TLM "allows women across the world to discuss anything and everything about what it means to be a woman in technology." In England, in 2013, Merici Vinton founded Ada's List, named after Ada Lovelace, the world's first computer programmer.

LIGHTNING ROD

Despite the benefits that accrue as you build a network, any effort to change the world will eventually be criticized, or even chastised.

Rachel was once quoted in the *Wall Street Journal* saying that things had to change to the point that, when people saw the "lineup for [major technology conference] TechCrunch Disrupt, they won't be able to not see the overwhelming maleness of it."

Michael Arrington, founder of TechCrunch Disrupt, used his platform to respond pointedly, telling Rachel to "stop blaming men" and offering the all-too-familiar explanation that all the women they tried to get on the panel said no.[6] He added, "[R]ealize this—there are women like Sklar who complain about how there are too few women in tech, and then there are women just who [*sic*] go out and start companies. Let's have less of the former and more of the latter, please."

Using facts, figures, and quite a bit of irreverent humor, Rachel replied to this criticism by making her case on Twitter. Arrington, in turn, invited her to talk about women in tech on a panel at the September 2010 TechCrunch event. The seven-person, all-female panel would be led by a woman, reporter Sarah Lacy.

Sklar describes the incident as "an ambush," which is confirmed by the video transcript.[7] "I don't even think this panel is necessary," began Lacy, who then turned to Rachel and inaccurately said, "Essentially we're here because you think TechCrunch is holding down women. That's what started all this."

"So, instead of talking about how to change the ratio, or even that it should be changed, they [tried to] turn me into the villain," Sklar recalls. "I got called an interloper and a self-aggrandizing person." Things only got worse from there, and the messy and poorly organized panel was later labeled by the press as a "catfight," which reinforced a common stereotype of what takes place when multiple women are involved in a debate.

As you find those like you, be aware that those *unlike you* are going to challenge you.* This is not meant to scare, but to prepare you.

---

* The usual reaction to a dissenter is negative. They are seen as annoying, wrong, and unpredictable. They are not "team players," yet these rogues are the ones who devise new solutions, per research done by Charlan Nemeth and Jack Goncalo: "Rogues and

FORGE A NEW PATH

Rachel was accused of being an interloper, which by definition is "a person who becomes involved in a place or situation where they are not wanted or are considered not to belong." To the audience she was disrupting—Silicon Valley titans who thought of tech as *their* space— Rachel *was* an interloper. While her lack of Silicon Valley–based tech industry knowledge and connections made her an outsider, they were also a source of her strength, as they enabled her to comment objectively. She didn't have to temper her remarks to avoid offending friends, and her income wasn't tied to that particular marketplace. Her standing within that particular group thus gave her both freedom and strength. Because she was observing from a safe enough distance, she was able to squint past the difference in specifics to spot a larger pattern, one that she had already observed in the media and in the political sphere.

The fact that what qualified her to chase her dent was also being used against her is an example of how an onlyness can have both a light and a dark side. It is onlyness, too, which supplies the answer for how to handle deflectors: Instead of denying it, claim it. If you can tell yourself a message that integrates the criticism, it takes away the shame of it. In Rachel's case, she was effectively about to say to herself, *My outsider status is what lets me see this with fresh eyes.* Integrating or reconciling the accusations, rather than getting distracted by them, will free you to do the work you want to do.

To create change, you obviously have to challenge the status quo by making an argument that may not have been made before. Which is not to say that you will get everything right initially; as you're breaking new ground, your critics will use mistakes you make as ammunition against you. "You don't know enough" will be their first line of defense,

Heroes: Finding Value in Dissent," in *Rebels in Groups*, ed. Jolanda Jetten and Matthew J. Hornsey (Malden, MA: Wiley-Blackwell, 2011), 17–35, http://onlinelibrary.wiley.com /doi/10.1002/9781444390841.ch2/summary.

and they will use their authority and position to disparage you. They might even attack you personally before they consider your point of view.

And in some ways, their attacks are not entirely out of line. After all, it's not easy to discern between the self-promoter who's just seeking attention and the groundbreaking, status quo–changing person with a bold idea. I mean, *you* know what's in your heart, and what only you see that needs changing, but how could *they*? Research shows that the attributes of high-trust leaders—that of dynamism, inspiration, articulation, and so on—can *also* be the attributes of self-serving, self-aggrandizing, narcissistic people. (Really!) It's the difference between character and charisma.*

So how do you deal with all this? What is it that you do that helps people tell if what they're witnessing is strong dent-making character or narcissistic charisma? At first glance, not much, because words fall flat. (Remember, the self-aggrandizer can talk just as well as, and sometimes better than, the change agent.) Which is the reason that people will wait, to see how you shake out. If you don't acknowledge that this is a real concern, you're not going to understand why people are resisting you at first.

And how do any of us know if a person has good character? The most important way is actions. We watch what you do piece by piece, what you do over time. A champion's behavior speaks loudly; over time, participants can evaluate what is going on. Knowing this, the best way to debunk the question of whether you are self-serving is by

---

* Charismatic leadership requires dynamism, image, inspiration, impression management, and so on. But those same traits can also be used by narcissists or authoritarian leaders—those who are self-serving, self-aggrandizing, and exploitative of others. Thus, Yassin Sankar, a professor of management in Nova Scotia, has found in his research that the key difference is the character of the person. Charismatic leaders with strong character are more likely to emphasize the mission rather than themselves and seek internalization of the idea (instead of personal identification back to the leader) by many. See Y. Sankar, "Character Not Charisma Is the Critical Measure of Leadership Excellence," *Journal of Leadership and Organizational Studies* 9, no. 4 (2003): 45–55, http://journals.sagepub .com/doi/abs/10.1177/107179190300900404.

letting your actions show your unwavering commitment to the cause over your own self-interest.*

It's your deep commitment to your purpose, grounded in your only-ness, that will ultimately get you through. Keep pursuing what matters, for the long term. If an idea or cause or purpose is important to you, it also needs to be beyond you. You nurture it by enabling *others* to belong and co-own it, and maybe even turn that idea into reality. They might even do it better than you, so let them, because that's why you started on this road in the first place. Your responsibility is to demonstrate your integrity with respect to the idea by every action you take to support it.† And that includes letting that idea be owned by many.

---

* Power without discipline is often directed toward personal aggrandizement, not toward the benefit of the cause. For more about how power is needed to achieve big dents (but needs to be directed), see David C. McClelland and David H. Burnham, "Power Is the Great Motivator," *Harvard Business Review*, January 2003, https://hbr.org/2003/01/power-is-the-great-motivator.

† The word "integrity" is derived from the Latin *integer*, meaning "wholeness." It is defined as soundness of and adherence to a set of principles.

CHAPTER 5

Common Purpose, Not Commonalities

*People are free when they belong to a community, active in
fulfilling some unfulfilled, perhaps unrealized purpose.*

—D. H. LAWRENCE

THE BLACK LIST: REMAKING HOLLYWOOD

"A lonely guy buys an anatomically correct sex doll over the Internet
and then asks everyone to treat it like a real-life girlfriend. You should
read this script *immediately!*"

Imagine that you are a junior person whose job it is to read and vet
Hollywood scripts and that you have to walk into your boss's office
and sell *that* wild idea.

But with the dent that Franklin Leonard started, many people in
Hollywood are now able to make such stories—stories that reflect a
full range of human experiences. If you've seen *Slumdog Millionaire*,
The King's Speech, *Juno*, or *Selma*, you've already experienced some of
the results of that dent. The script of *Lars and the Real Girl*—the idea
that opens this chapter—went into production shortly after Franklin

began his crusade. Each of these highly successful projects had originally been passed over by Hollywood, destined for the dustbin.

What Franklin is accomplishing in Hollywood shows us a way to be both distinctly ourselves *and* deeply connected with others—not by fitting in but by being connected in a way that requires neither commonality ("we are all exactly the same") nor conformity ("we give up a part of ourselves to fit in"). The uniting power here is common purpose. When purpose is the glue, then the group's "us" does not suppress individual ideas at the expense of greater unity, and as a result, those ideas can grow big enough to dent the world. The way to achieve this is through defining (or, rather, *redefining*) *who we are to one another.*

THE WALLED CITY

Writers always approached Franklin at conferences with the same question: "I've written a great script, but I don't know anyone in the industry. How can I get it to the right people?"

His stock answer was both clear and discouraging; "Pick up your life," he'd say, "and move to Los Angeles, and while you work at Starbucks, network your way into knowing somebody."

That was, in fact, more or less how he had landed his first Hollywood job. After growing up in the south as a "nerdy kid," he'd worked hard enough to get himself admitted to Harvard University, then hopscotched through a series of several interesting but random jobs. Wondering if he'd ever find a real calling, Franklin found himself binge-watching movies one weekend and it finally struck him that he'd always loved film.

He moved to Los Angeles and hustled himself into a job at Creative Artists Agency. Like most entry-level jobs, this one was primarily secretarial: answering phones, handling correspondence, setting up meetings for his boss and her clients, and, at times, fetching coffee from Starbucks.

While working at CAA, he learned that Hollywood is effectively a walled city, completely closed to outsiders. Whether you were "in" or "out" affected everything in the movie business, from who got employed to what script ideas were considered. A few years and several jobs later, Franklin had reached the level of development executive, at which he was tasked with finding new, interesting scripts. Many, he began to realize, never had the chance to reach him, very often due simply to the writer's location or lack of connections.

"If you are a single mother with a mortgage and two kids, working, say, in the middle of Macon, Georgia, you can't pick up your life to move to Los Angeles," he explains. "Even if you're a phenomenal writer with a novel screenplay, the chances of someone in Hollywood seeing that idea [are minimal]."

Of course, with billions of dollars in annual revenues, Hollywood was hugely successful, but Franklin believed the industry could do even better, both financially and creatively, if it could make more stories that reflected the lives of all types of people. As he observed, "Studios are always trying to reverse-engineer a hit . . . into a core set of narrative flows, or heroic arcs. But stories or art . . . don't work like that. You can't be formulaic, because that's when the story doesn't touch us and reach a deeper truth."

What happens in Hollywood is not unusual in business. To run a hospital "efficiently," health care professionals follow standard protocols under which actual care becomes secondary. Educators are required to follow established standards, which restrict their using their own abilities to elicit a love of learning. People seek careers to pursue their interests and express their creativity, but in the rigid constraints of any industry's "system," anyone can lose the very best of himself.

Franklin, for his part, wondered if it was even possible for Hollywood to produce more films that spoke to him—credible narratives about underdogs, which is what he loved about movies in the first place. As he sees it, "Popular entertainment serves much the same role that religion once did . . . I don't think it's a coincidence how we feel afterwards—connected—after a great film. Movies are the most immersive, richest form of storytelling today, they help us know what it means to be human. They're basically modern morality tales."

FILTERING OUT, NOT FUNNELING UP

Hollywood decision makers typically receive such a vast number of scripts that they use a scheme called "coverage" to make the sheer volume manageable. Unpaid or low-paid interns, usually recent film school graduates with minimal industry experience, read the scripts and write a synopsis that highlights the plot, characters, tone, and genre of the story. Thus, the initial assessment is delegated to the business's least experienced individuals, those with low social capital and the least power. This process effectively *filters out* most new and novel ideas or perspectives rather than *funneling in* good ones. Good films still get made, obviously, but Hollywood seems to produce them in spite of its system, not because of it. The film industry seems to prefer the comfort of proven formulas to the risk of "it's never been done before."

Sitting in his office on Sunset Boulevard one day, Franklin Leonard composed an e-mail that would change how Hollywood finds great scripts. "It wasn't that I thought sending this e-mail was such a great idea, revolutionary, or *anything*," he explains. "I was just so desperate to find good screenplays that I acted. I was trying to go on vacation for two weeks with better stories because I had read a bunch of garbage for months."

Sent from a newly created Gmail account, blacklist2000, the message was an anonymous request to seventy-five people to share information. To build the distribution list, Franklin had pored over his calendar from the prior year, capturing everyone he had had meals or drinks with or he thought could be helpful. The e-mail asked:

"Send a list of up to ten of your favorite scripts from this year that meet the following criteria:

1: you love the script

2: you found out about the script this year, and

3: it won't be in production by the end of this calendar year."

In exchange, he promised to share the combined list.

"It was a very direct, quid pro quo transaction," Leonard recalls. "Though I wasn't counting on everybody participating, pretty much everyone . . . responded and said, 'Hey, I've also got a friend who would love to participate.'" Most of the recipients didn't even know the e-mail came from Franklin.

By the following day, ninety-three people had submitted their anonymous responses.* Franklin got to work, collating the responses into a spreadsheet pivot table, counting each nomination equally. He saved the results to a PDF file named "Black List,"† a cheeky nod to the witch hunts that ruined so many Hollywood careers in the 1950s.

"Black List" also drew on Franklin's history and experiences as an African American, his visions, and his hopes—in other words, his onlyness: "In an English class in middle school in south Georgia, I remember exactly when the teacher was explaining color symbolism in literature. *Shane* was the book being used. They were explaining it as, the cowboy wearing a white hat, they're probably the good guy . . . and I remember sitting there thinking, 'Wait, whoa, I *do not like* where this particular story line is going.'" Once he was taught that the black hats were probably bad guys, Franklin remembers envisioning that he would one day "write a book that completely inverts this color symbolism."

Ultimately, he did reframe the color symbolism of "black," but in a better way than just swapping one foolish convention for another.

---

*   Anonymity can reduce personal accountability and impede independent and equal contribution, which is a central attribute in making the "wisdom of crowds" work. But in the case of the Black List, anonymity actually fostered independence and integrity by allowing contributors not to be concerned about potential conflict, threats to their status, or offending the existing power structures.

†   You can see the first year's list in the archive: http://files.blcklst.com/2005_black_list.pdf.

After sending out the PDF, Franklin printed out some of the scripts that had been submitted and went on vacation. In his absence, the Black List went viral. "By the time I got back from my break," he explains, "the Black List PDF had been shared back to me via e-mail more than a hundred times—because I didn't put my name on the list, people didn't know . . . that I was the original creator."

Many of the recipients—some of them "powerless peons," in Hollywood terms—took the list to their bosses and repitched ideas that they couldn't sell before. Some of the scripts that had been previously acquired but had then been shelved because they were deemed not commercial enough were reevaluated and went into production.

Unrepresented writers, who had no other way into the system, were now having their novel stories considered. Hollywood, like many industries, believed that its system was a meritocracy, and that anything truly worthwhile would rise to the top. But they did not take into account the degree to which their institutionalized biases were limiting what they considered. The Black List radically enabled ideas to be evaluated based more on true merit by adjusting for the bias. Franklin saw how this challenged the status quo, and he was thrilled.

ARE YOU BEING A REBEL OR A LEADER?

Franklin once made himself a T-shirt with bold iron-on letters that read TOKEN. He'd wear it to parties when he knew he'd be the only black person in attendance, waiting for a reaction. "I would see some people get it," he says, "and think it was the funniest thing in the world. Others would get really offended by it. And some others just wouldn't get it at all. It was always something that I enjoyed playing with."

Franklin's shirt was a provocation, while the Black List applied his rebelliousness to a greater purpose: to lead with an idea.

To rebel is to push *against*; to lead is to advocate *for*. To rebel is to say "we won't"; to lead is to say "we will." To rebel is to deny the authority of others; to lead is to invoke your own authority. Rebels attack, while leaders drive toward something.* Franklin Leonard, through the Black List, is acting as *both* a rebel and a leader. The word "leader" is used in this context to describe not just those who have assumed a formal mantle of leadership but anyone making a dent—pushing for solutions, agitating toward progress, and making things happen. While it's possible to be a rebel without being a leader, you can rarely be a leader who achieves a dent without also being a rebel.

The two roles of rebel and leader can be reconciled in the form of the *protagonist*. A protagonist is a principal champion of a cause or program or action. The protagonist does not wait for permission to lead, innovate, or strategize but does what is needed and right for the group as a whole, without regard to his or her own status. Just as in Hollywood films, without the triumph of a protagonist, the story lines of industries, businesses, political movements, or any group that seeks to bring about innovation or change can never achieve a satisfying resolution. The protagonist makes a transformative dent by believing in possibility and then acts on this belief by getting other people engaged toward specific new outcomes.

OUTSIDERS ENTER THE WALLED CITY

At the time he started the Black List, Franklin didn't want to be recognized for his work and told only close friends of his role in the project, in part because he valued his development executive job at Leonardo DiCaprio's firm and feared losing it. People might also question his ability to find good scripts if he had to ask for help in obtaining them.

About six months after the Black List debuted, Franklin got a surprising call from one of the powerhouse agencies in the industry. It was

---

* I wrote about these ideas in the *Harvard Business Review* ("Are You a Rebel or a Leader?" January 25, 2011, https://hbr.org/2011/01/are-you-a-rebel-or-a-leader). To be a champion, you can't just be that person who says "this is broken." As Franklin shows us and we'll see again in chapter 7, raising a ruckus and creating resentment is less productive; achieving new outcomes is what counts.

a run-of-the-mill pitch until the agent, in a secretive voice, added: "Listen, I have it on really good authority that this is going to be the number-one script on next year's Black List." Franklin laughs as he recounts this story. Behind him in his Los Angeles bungalow is a wall of bookshelves groaning under the weight of scripts. Franklin explains the irony: "I hadn't even decided if I wasn't going to do the Black List again."

He understatedly refers to this moment as "a clue."

By 2015 he had issued a Black List for ten years, with staggering results both artistically and financially. Nearly three hundred of the one thousand Black List scripts have been produced, earning over $25 billion worldwide. They've received 223 Academy Award nominations and won 43 Oscars. Four of the past six Best Picture winners, ten of the last fourteen screenwriting winners, and three of 2014's screenwriting nominees were Black List scripts.

Most interestingly, for each of the first eight years, the Black List's top five scripts were submitted by "outsiders"—writers not living in Los Angeles nor represented in the industry. The Black List was opening doors in the walled city, and new people and their ideas were coming in.

WHAT CAUSED THAT SUCCESS?

"I can cite all the numbers about the successes of the movies of the scripts that have been on [the Black List]," Franklin says, "but I don't want to take credit for those movies getting made, or those movies getting made well. That credit goes to the people who make the movies. I think what we do is shine an incredibly bright spotlight on the projects that deserve it. The consequence of that very bright spotlight is to catalyze a lot of these projects toward or into production."

While the Black List did create a spotlight and used the wisdom of crowds to make a dent, that's not the most useful part of its story. It incorrectly suggests that someone could simply fund a big enough spotlight. But money and other traditional forms of power wouldn't accomplish the change the Black List did. What actually worked was that Franklin got the group to rearchitect the relationship of *who they are to one another.*

ASKING A NEW QUESTION

To understand how that process worked with the Black List, it's worth examining it in more detail.

When Franklin described the e-mail as asking for favorite scripts, he never slowed down to emphasize what I consider to be the most critical part: "What screenplays do you most love?" To understand the significance, note what he didn't ask, "What do you think will work in the marketplace?" or "What will make the most money?" If he had made his pitch in those terms, nothing would have changed, because precisely those blockbuster-oriented questions were the existing scheme that Hollywood was already using to filter scripts. "For years, there was a conventional wisdom about the idea that you can't sell female-driven action movies," says Franklin. "I remember when I was working at Universal, I read *The Hunger Games* . . . and said, 'We should buy this.'" He kept trying, but was told, "Doesn't work. Leave it alone." Color Force, an independent studio founded by producer Nina Jacobson, eventually bought the property and coproduced the film with Lionsgate Entertainment, and the project was an enormous success.

The same sort of conventional wisdom that led movie studios to reject female-driven action films also held true for plots involving people of color. "There's an assumption that you can't sell black actors abroad," Franklin explains. "When you say, 'What about Will Smith?' [the response is,] 'Well, that's the exception, which proves the rule.' 'Well, what about the fact that *Coming to America* made $300 million foreign when it came out?' 'Well, that was Eddie Murphy, that's a different thing.' Just how many assumptions do you need to disprove before you admit that maybe the model itself is just wrong?"

Every industry has a set of accepted beliefs about "what works," which typically gets passed on as immutable gospel. In business, disruptive ideas are often barricaded or weeded out either because "That's not how we do it around here" or because "If it was a really good idea, we'd have heard about it by now," or by dismissing the disruptor herself. But asking questions that contest the status quo can invite a complete reexamination of existing norms.

It can point a group toward a more compelling direction.

New questions challenge the invisible walls that frame an idea. Framing is the cognitive structure that guides what happens to new experiences and ideas. These filters act as a cognitive sieve, saying *this* is what matters and letting that stuff go through, and keeping all the other stuff caught in the sieve, left behind. To be fair, most of us don't realize or notice or care what frameworks are at play in us. But by recognizing frameworks for what they are, you can change them. By asking a new question, that's what Franklin is doing: recognizing that frameworks that have been conceived can also be reconceived,* and that doing so can completely change the conversation at hand.

Finally free to speak and voice their own views, the script readers changed the game. Novel scripts began to be funneled in instead of filtered out.

THE US IN OUR DREAMS

The core tension of being a part of an "us" often comes down to the dilemma: "Do I have to give up *me* to belong to *us*?" On a personal level, it appears in the dynamics of couples, in that moment when one partner might wonder if staying in the relationship will require surrendering on an important point to keep the peace. In teams, it can

---

* Gail Fairhurst, a professor at the University of Cincinnati, has written several influential books and papers on framing, which informs this line of thinking. Fairhurst reminds us that the world we live in today is conceived and framed in a particular way. This shapes our experience. Even the language we use orders and reorders social life. The Old Guard, and those in power (and for Americans, this can be defined by the cultural norms, which are heterosexual, old, white, male, and rich), don't even recognize this issue of frame. It's just "the way things are."

occur when you realize nobody is listening to any of your creative ideas because of your title, so you remain in the job but stop contributing. If our ability to express our ideas gets us kicked out when we have the need to stay, we're caught in a bind. So we give up on *ourselves* to belong to *us*. The question is, how do you serve *both* your own interests and the group's to create and thrive? It's not by saying "it's you against me" but by figuring out how we face a common problem, framed in such a way that it unites us.

The power of *us* is our common purpose, when it defines an "us" that all its members aspire to be a part of. Simply because we're accustomed to a known system doesn't mean we can't change it for the better by asking a better question.

WHEN MEANNESS RULES

In some cases, however, becoming part of a committed group can have negative consequences. When people are united around a noble purpose, they can create a compelling "us." But they can also join forces for reasons that are far from altruistic.

For two weeks in March 2007, Kathy Sierra, a programming instructor and developer, found herself on the receiving end of the Web's dark side and was forced to go into hiding after receiving threatening blog posts and e-mails.[1] One message included a picture of her head in a noose; another was a threat to rape her by sundown. Trolls even made her home address public.

"I'm afraid to leave my yard, I will never feel the same, I will never *be* the same," she wrote on her personal blog.[2]

What could have prompted such a horrendous attack? Surprisingly, it was a response to her blog about Web design, Creating Passionate Users,[*] on which her smart, warm content included how people's needs mattered in tech. Her onlyness was based in cognitive science research,[†] motivated by epilepsy early in her life. Technorati, an industry website, ranked her site as one of the top fifty tech blogs.

A few weeks earlier, Kathy had written that sites like hers should be able to delete user comments, and that Web dialogue shouldn't be just a free-for-all but should have guidelines. Some of her readers didn't agree, and to show her their displeasure they posted hateful notes on her blog and another (now defunct) site, Mean Kids. The "protest," as the anonymous users called it, was forcing Kathy to deal with hate in order to do something she loved.

When groups of online users become mobs, the Internet's power is

---

[*] The site remains up and archived as a resource for people; comments are now shut off. Take a look, here: http://headrush.typepad.com/creating_passionate_users/.

[†] Kathy Sierra, "Who's in Charge—You or Your Brain?" *Creating Passionate Users* (blog), April 11, 2005, accessed February 21, 2017, http://headrush.typepad.com/creating_passionate_users/2005/04/whos_in_chargey.html.

used for ill. The same tools that enable communities to create meaning can also be a weapon of social malice. Online harassment disproportionately affects women, people of color, and LGBT individuals.*
But such cruelty also limits the dissemination of new ideas and
stops the people associated with them from ever having a chance to
share them.

Online feuds seem to have become just part and parcel of Internet
culture. But the problem has clearly escalated to the point of being
destructive. In 2003, Clay Shirky described how online groups could
be their own worst enemy,[3] arguing that the many-to-many aspect of
Web-enabled communication can transform a group into a mob in a
very short time. But no one in the tech community, law enforcement,
or advertising has taken responsibility for addressing this.

THIS COULD BE YOU

Kathy Sierra wound up shutting down her blog, and we collectively lost
a fresh, strong voice in technology. Her silence meant that we no longer
had access to her ideas or their ability to make our own ideas better.

Pew Research has reported that 39 percent of people online have
experienced bullying, online harassment, and intimidation.[4] Some 70
percent of users between ages eighteen and twenty-four say they've
been the targets of harassers. In 2013, a man posted phony ads supposedly announcing that his ex-wife was soliciting sex.[5] One was titled
"Rape Me and My Daughters" and included his ex-wife's home address.
More than fifty men showed up at the victim's house. Tyler Clementi,
an eighteen-year-old Rutgers University student, jumped to his death
from New York's George Washington Bridge in 2010 after his room-
mate posted a secretly recorded video of him kissing another man.[6]

---

* For people who aren't hetero white men, the Internet can be a threatening place. In 2015,
  Wired magazine convened six people to discuss how to take back the Net: Laura Hudson,
  "6 Experts on How Silicon Valley Can Solve Online Harassment," *Wired*, October 22,
  2015, https://www.wired.com/2015/10/how-silicon-valley-can-solve-online-diversity-and
  -harassment.

Justine Sacco, who had only 170 Twitter followers, tweeted an ill-considered "joke" about getting AIDS just before boarding a flight to Africa.[7] By the time she landed, the comment had gone viral, she had been publicly shamed, and she had lost her job as a director of corporate communications.

We can all be victims of online bullying; there is no protected class. If there is no safety for each of us, there is no safety for *any* of us.[8]

HATE IS HATE

Almost a year after the Kathy Sierra incident, the *New York Times* ran an article titled "'The Trolls Among Us." The piece identified Andrew Auernheimer, who goes by "weev" online, as the person who had shared Sierra's home address and Social Security number as well as degrading fictional claims about her. There were no consequences to his doing so, as such behavior is often defended in public or in court as free speech.

In a society where we enthusiastically embrace free speech as a right, we often fail to take into account that with every right comes a responsibility.

Brianna Wu, a video game developer (and cofounder of Giant Spacekat, an independent video game development studio), was the target of one of the most public incidents of online bullying in "Gamergate," a controversy that is in its third year at the time of this writing. The gaming industry was once dominated by males, and almost all its customers were men. As women began playing and creating games, they threatened the status quo. Wu describes the groups that have attacked her as the "KKK of the gaming industry."

She's naming what all this is: hate.

Reddit, dubbed by some as the front page of the Internet, has deferred to the crowd on the issue of free speech. The result is that the Southern Poverty Law Center has described it as the number-one spot for violent racists: "The world of online hate, long dominated by website forums like Stormfront and its smaller neo-Nazi rival Vanguard

News Network (VNN), has found a new—and wildly popular—home on the Internet. Reddit boasts the ninth highest Alexa Internet traffic ranking in the United States and the thirty-sixth worldwide. Many of Reddit's racist subreddits are among its most popular."*[9]

Jason Pontin, editor in chief and publisher of *MIT Technology Review*, has written that the position of online activists is that vigilantism and censorship is good when it "punches up"—meaning when it challenges the powerful. But, Pontin argues, a more accurate view would be that "such 'activists' are not principled, neutral defenders of free speech or civility" but, rather, "proponents of group and class interests."[10]

This meanness is collectively suppressing those who will disrupt the status quo.

IT IS SOLVABLE

In 2011, Internet entrepreneur Anil Dash warned that we're allowing meanness to win by not using what knowledge we have.

"This is a solved problem," Anil states. "We have a way to prevent gangs of humans from acting like savage packs of animals. But the online world is just ignoring most of the lessons we've gathered over the last thousand or so years from disciplines like urban planning, zoning regulations and crowd control." And so, he argues, "If you could apply these sets of principles, you could prevent the overwhelming majority of the worst behaviors on the Internet."[11] The examples he cites are having real humans dedicated to monitoring and responding to your community, establishing community policies about what is and isn't acceptable behavior, making your site have accountable identities, implementing the technology to easily identify and stop bad behaviors, and planning for a budget that supports a good community. Applying those principles would not require revolutionary action; im-

---

*   As this book goes to press, Reddit seems to have improved policies, shutting down some nutso subreddits; see Luke Lancaster, "Reddit Shuts Down 'Alt-right' Subreddit," *CNET*, February 1, 2017, https://www.cnet.com/news/reddit-bans-alt-right-subreddit.

plementing simple policies can go a long way to prevent online interactions from spiraling into malicious gang attacks. The broad lack of such clear policies makes the Web fertile ground to oppress people.

Anil's key point is that if you're allowing such behavior on your site, it's your fault. We know that a reasonable degree of control is possible, such as when, during the 2016 Olympic Games, Twitter was able to determine almost instantaneously who was posting illegal videos and take them down—proving that it's not about the social software but the social norms that are allowed and enabled. If they can do it for the Olympics, surely they can also allocate some resources to making the Web a safer place.

So where does this leave us?

Let's start where we can. "Be an upstander, not a bystander," as Monica Lewinsky said during her TED talk regarding her role as Patient Zero in the runaway epidemic of online mob behavior.* Each of us has the ability to stand up to bullying, online and off, even if we don't have a stake in a particular conflict. Be a human firewall by stepping forward with a post, and ask the hateful person to step it back. While bystanders may seem as if they're simply not taking a stand, in reality their neutrality amounts to tacit support for whomever and whatever is the status quo. Upstanders do not lurk indifferently on the sidelines. In this battle we are actually in a struggle for how to be ourselves, with each other. The future is not predetermined. It's up to us to make the choices that will drive the healthy Internet culture we believe in.

---

* Twenty years ago, Monica Lewinsky, a young intern who had fallen in love with her much older, powerful boss (the president of the United States), found herself and her personal story splashed across the front page of the *New York Times*. Since then, she has used her history and experience, vision, and hopes to raise the issue of bullying, to turn her dark experience into a light for others. Jon Ronson, who has written an entire book about online bullying, wrote a story about Monica here: "Monica Lewinsky: 'The Shame Sticks to You Like Tar,'" *Guardian*, April 22, 2016, https://www.theguardian.com/technology/2016/apr/16/monica-lewinsky-shame-sticks-like-tar-jon-ronson.

CHALLENGING THE AGE-OLD TRADITION OF *SWARA*

After four-year-old Asma's uncle was found guilty of committing murder, she was offered up by her Pakistani tribal leaders to another family in the village as "compensation." Instead of imprisoning the criminal, the culture dictated that Asma would pay for her uncle's freedom with her own.

When Samar Minallah Khan learned of this practice, she used her video camera to capture and share stories like Asma's. But to make the dent she wanted—to prevent young girls from being traded by the male tribal leaders of the community—she had to join forces with people quite unlike her.

How do you unite parties with fundamentally deep differences? You can't strong-arm people into respecting your point of view, but you can add more information for them to consider. You can't discount others' opinions or diminish their traditions, but you can try to disturb the status-quo thinking. This is precisely what Samar did when she first learned of *swara* in 2002, from a man who used the term as a point of comparison: "It was like *swara*—when a girl is given, it is a symbol of peace."

As a trained anthropologist, with a master's degree from the University of Cambridge, Samar had always loved learning about tribal practices that were specific to particular regions. When she investigated *swara*, what she found out was deeply disturbing to her. The practice involves a young girl being given away from one family to another as restitution, typically for a murder or other serious crime. It has accounted for thousands or perhaps even hundreds of thousands of girls' fates. Though the name varies across different Muslim areas of Punjab, Pakistan, and Afghanistan—in Punjab, it's *vanni*; in Sindh, it's *sang chatti*; in Afghanistan, *ba'ad*—in every region, their futures are stripped from the girls before they reach puberty.

Those decisions are made by members of a council of local men,

called the *jirga*. They are the village elders and the ones who keep order, governing using the word of Islam called sharia, or sharia law.
Depending on how severe the crime, or how many girls the offending
family has, the council discerns exactly how many girls must be given
to the family who has suffered a loss. Sometimes a judgment can be
made but not acted on immediately; for example, if the girl is too
young, she'll be handed over when she matures.

Since this is a "reconciliation" custom, it is celebrated with a feast,
during which a lamb is ritualistically slaughtered, and sometimes the
entire village is invited. In the course of these festivities, which are
hosted by the party that is having its honor "restored," the *jirga* leader
shares the final judgment and tells the village exactly what will happen next. The murderer is then forgiven in the eyes of Islamic law and
by the village. He is free; the girl is not.

HATRED, NOT RECONCILIATION

Wanting to give voice to the voiceless, Samar decided to find and record
swara stories, because without specific accounts of the people whose
lives it ruined, *swara* would be interpreted as an abstract concept.

But to obtain the women's testimony would be difficult, as they
would never share their stories in the company of men outside their
immediate family or if it meant dishonoring their own family. Because
all the cameramen she knew and had access to were men, Samar explained, "I knew this meant I'd have to learn filmmaking to avoid
bringing a male cameraman into such a personal conversation." She
began her now long and successful documentary career by reading
Filmmaking for Dummies.

Her journey took her first to Khyber Pakhtunkhwa, an area in
northwest Pakistan that was bombed by America after the September
11 attacks. After four visits to different villages within the region, Samar was unable to get anyone to talk to her about *swara* and started to
wonder if the tradition was merely a legend. "People would tell me it

didn't happen in their village, or that it didn't happen anymore—that it was a thing of the past."

Finally, in the fifth village, she found a woman working alone in a mountaintop field in the Swat district, one of the Provincially Administered Tribal Areas of Pakistan. It was early evening, and the woman agreed to accept some tea from Samar and talk, as it gave her an excuse to rest after a long day of farming.

It was from her Samar learned that *swara* is very real.

"When I mentioned *swara* to her, she first shared her experience of the girls in the village," Samar recalled. She spoke to the woman in Pashto, the local language, which helped establish trust. After speaking for an hour or so, the woman finally shared that she had a personal story of *swara*, a deeply painful experience. Samar filmed her as she tearfully recounted how she had had no choice but to surrender her ten-year-old daughter to another family because of a crime her own father—the girl's grandfather—had committed.

What Samar recorded in this film vignette is testimony to a serious, yet largely undiscussed issue. UNICEF, which tracks child marriages, reports that 70 percent of girls in Pakistan are married before the legal age of eighteen, and 20 percent by the age of thirteen.[12] Though no solid estimates exist for how many of these marriages are specifically linked to *swara*, one doctor in Lahore who treats these girls when they have psychiatric breakdowns says a *swara* girl is traded every three seconds.[13]

In her documentary, titled *Swara (A Bridge over Troubled Water)*, Samar reveals the truth about this practice. A tribal chief from Khyber Pakhtunkhwa admits, "I might let a *swara* girl into my house, but I cannot let feelings of kindness to her in my heart." Even though *swara* is meant to end warring disputes, *swara* girls become the targets of all the residual anger and hatred related to the original crime. "They are often emotionally tortured and sometimes raped by other men in the family. They are made to suffer for a crime they did not commit," Samar explains.

How can someone bring change to such a pervasive but unacknowledged issue? This book has examined how to pursue one's passions, but what if those passions are inflamed by rage at a particular injustice? Samar's amazing gift is her ability to accept that each of us has onlyness that is worth honoring—even "the enemy." This empathy has enabled her to bridge a seemingly unbridgeable chasm of difference.

DISTURBING, NOT DIRECTING

Born to a Pashtun clan in Peshawar—one of the largest cities in Pakistan—Samar was fortunate to have a father who was, to use her words, "pro-women," and saw to it that she was as well educated as her brothers. Though she was raised with urban, progressive influences, Samar was familiar with more conservative, traditional village life, as she spent her childhood holidays in a village much like the ones she would later visit for her documentary.

"When I looked at these *swara* girls," she recalls, "I saw in their faces the girls I grew up with. Any of them could have been my childhood friends."

Because of her upbringing, Samar has always straddled the worlds of traditional Islamic culture and that of the West, so she knew from the first she'd never get people to understand *swara* if she used Western-type arguments or framed the problem as a "human rights issue." Since *swara* was intricately tied to cultural and religious customs, any change to the practice would have to be tethered to that framework. She would have to find the answer by finding and building a common ground between tradition and the girls' perspectives.

A devout Muslim, she turned to the Quran, where she read that "whoever commits a crime, it is that person [who] needs to be punished for the crime." With this verse in hand, she began reaching out to religious scholars and local imams in different parts of Pakistan. She knew that she would have to determine whether *swara* was actually mandated by religion, and that the answer would have to come from individuals the tribal leaders would regard as authoritative. Samar accordingly filmed experts interpreting this verse and discussing the teachings of Islam as they related to *swara*, so that their voices could be used to influence the *jirgas*.

"At times, even [the priests] don't know about that particular verse because we [Islamic people] don't read the Quran with translation." Culturally, in Islam, the holy text of the Quran is read in its original Arabic, which would be akin to reading the Bible in its original language, even if you don't know Hebrew.

One of Samar's meetings in Islamabad led to an introduction to Dr. Fida Muhammad Khan, a sitting judge on Federal Shariat Court, who stated that *swara* was actually being misused by the local *jirgas*, as "*Swara* has no standing in Islam."

Dr. Khan explained his interpretation in Samar's film: "There are three laws to address a murder. There is the death penalty, or to give what is called blood money. Or a third option: which is a compromise with the intent to make peace. Sometimes *swara* is explained under this third way." But, Dr. Khan clarified, "according to Islamic law, there can never be vicarious liability," which means no innocent lives should be sacrificed for the crimes of others.

"This [*swara*] is against the teachings of Islam," Dr. Khan concluded. "It is recorded that a young woman had come to Prophet Mohammad to say her father had forced her to marry, and the prophet said this is wrong."

Samar also traced the origins of *swara* to old literature. "In the past, this custom was not even practiced the way it is being practiced today. It was more as a symbolic custom. A girl would be sent to an enemy's family, but there she would be presented with a chador [a veil, which is a sign of purity and respect], given gifts of sweets, and then she would be sent back to her parents' home. The fact that a girl was sent to an enemy's family, that in itself was seen as enough [to seek forgiveness], in the past."

How Samar went about preparing to implement her dent is instructive. Unless people work directly for you, or you can make them do something through laws or fiat, you cannot direct anyone else—let alone a group—to change.* Convincing a group to do your bidding simply because you demand it is not actually a useful *change*, it's simply *capitulation*. When you are seeking capitulation, you are simply trying to invert the power structure that has repressed your ideas. But doing so is not only unhelpful, it's not onlyness-fueled power. What you're actually conveying by seeking capitulation is that *your* expression of "only" is the *only* one that counts.

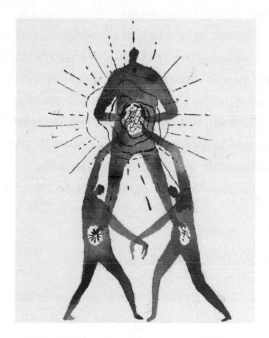

One Onlyness Cannot Dominate over Others

---

* In a market economy, you can direct people by having courts and other arbiters stipulate who wins and who loses. And in hierarchical constructs like organizations, you can tell people what to do because someone sets directions and then supervises others to make sure they do what is directed. But in the network-power context we're talking about— the world where connected people gather together to make a dent—you can't direct people to act. As onlyness is clearly drawing on a network theory of power, useful reading on that topic is Manuel Castells's research: "A Network Theory of Power," *International Journal of Communication* 5 (2011): 773–87, https://faculty.georgetown.edu/irvinem /theory/Castells-Network-Power-2011.pdf.

While getting others to follow your path in this manner might *feel* like a win, it never results in a dent being made, because it is not based on the commitment that comes from within, from a shared purpose, from honoring each of the members of the groups involved.

Without realizing it at the time, Samar was setting up the conditions to *disturb* the status quo. As Humberto Maturana and Francisco Varela, two Chilean philosophers who study how humans operate, explain, "You can never direct another living system. You can only disturb it."[14]

New information can be disturbing, especially when it comes from a respected source—and even more so if it concerns revered matters like tradition, culture, and religious direction. To disturb someone while honoring onlyness is to challenge the idea without challenging the worth of the people involved. By providing more content and perspectives, Samar made it easier to debate the ideas around *swara*. In such a debate, the ensuing struggle is not, then, a battle between egos but a battle *inside* each person to resolve or reconcile the inconsistency of the new information with their prior decisions. This is the key to change. The information itself doesn't create the change, the tension does.

THE SUPREME COURT AGREES

Before she gained the support of the *jirgas,* Samar was able to get assistance from the press. In 2003, her first film on *swara* practices was screened at the Peshawar Press Club, access that her stature as the daughter of a well-known bureaucrat surely helped her gain. The movie itself caused a stir, challenging the existing assumption among urban dwellers that this kind of practice no longer took place. The film featured a poignant scene of a man leading his five-year-old daughter to a *jirga* council to offer her up for *swara*, an image that was repeated several times throughout the film. As a reviewer noted, "It is as chilling the last time as when it first appears."[15]

Because of the press influence, the film was widely screened and shared, finally reaching the highest echelons of the Pakistani political system. As there was no existing law forbidding *swara*, Samar insisted that the time had come for it to be declared illegal. In 2003, she filed what is called a "public litigation plea" in Pakistan's Supreme Court,[16] asking it to end *swara*. Her filing led the courts to direct the police of all four provinces of Pakistan to halt the practice of using marriages to settle family feuds. During that year, Samar also reached out to parliamentarians, government officials, and civil-society organizations to lobby for legislation. In 2004, the penal code PPC-310-A made *swara* a crime punishable by up to ten years in prison. In 2005, 180 *jirga* decisions involving *swara* girls were prevented by local police.

BUT EVEN EDICTS DON'T CREATE CHANGE

Even though the Pakistani penal code was changed, the practice of *swara* continued, because it was the *jirgas* who governed the local regions. "The law criminalized such [*swara*] practices, but the *jirgas* remained a major obstacle in implementing it," explained former parliamentarian Dr. Attiya Inayatullah. She advocated that the *jirga* system be abolished, arguing that "the *jirga* is not in the Constitution or in Islamic law, and it is time for a law to remove women and children topics from the domain of the *jirga*. You cannot have a parallel legal system that can do what it wants, and especially on decisions of this nature."

Samar herself had come to realize that if genuine long-term change was going to happen, it was going to take something more than a power struggle over who gets to decide. "I had gone from being a curious anthropologist to this angry activist who thought that these tribal councils and the Pakistani village men are simply evil because of what they are doing. I would argue with any father who wanted to give his three- or four- or eight-year-old girl in compensation. I would become this crazy woman who didn't know how to explain to any father that 'what you're doing is wrong.'" But, she admits, she had to slowly learn to stop responding with anger, "because it clearly didn't work."

While anger rarely unites people productively, it does serve an important role: It is the sound of your values screaming to you. You need to listen to the anger to hear what matters to you, but externalizing it and directing it at others is a trap. To talk in tones of anger is to enter into a dialogue of the deaf—much like my mother and I did—in which neither side will hear the other.

Samar's first village visit after the anti-*swara* law passed became a transformative moment. Unexpectedly, she discovered that the tribal council was about to decide whether an eight-year-old girl, Marina, was to be awarded as compensation for her uncle's having committed a heinous crime. Samar met with the girl's father over tea to explain that the practice was no longer legal and to share her perspective and what she'd learned in the Quran.

The man's reaction—and those of others she later met—surprised her. "Some of the men I reached out to looked like a typical bearded Taliban man that you see on the Western media," she recalled. "The difference was that the ones I came across would have tears in their eyes and empathy in their heart for their daughters." One she says she can never forget; his voice was trembling as he said, "I would give anything—my land, anything [to save my daughter]."

Having experienced the Taliban to that point only as a group, Samar hadn't been aware that there were individuals among them who were ready to stand up against such customs. For her, this was a moment of empathy, a moment of realization that sometimes people can simply be ensnared in history, doing what *has* to be done. If you see such individuals as a "them," or only through the lens of the group, you deny them their own unique story, their onlyness. For Samar, the issue became far less black-and-white; it stopped being "Samar vs. the Tribal Council" or "Samar vs. Fathers."

With the grieving father's support, she was able to attend the *jirga* meeting where the decision about his daughter would be made. She was dressed in a traditional white *chador* (veil) from head to toe, so that her face was all that the *jirga* could see of her.

"They agreed to let me sit there," Samar recounts. "But then, when I spoke respectfully to them to thank them for letting me sit and listen, they even let me speak to them. I told them that this is how the Quran says that this is un-Islamic, according to this devout expert. This is what our [new] government law now says. I talked to them about how this was practiced in the past—on how I know culture is such a great thing, but it was just a symbolic kind of a custom. I didn't tell them what to do, but just gave them these options of how Pashtuns would do it in the past—the *real* Pashtuns. That let me share the other options, such as a piece of land, or turning over a home, anything other than giving of a girl who will face a terrible life for a crime she did not commit.

"All I want from you," Samar announced, "is to share my thoughts with you."

She then showed a film of a father from a tribal area, letting his words speak for themselves. When it concluded, she left the council room, as it was the custom to allow the *jirga* to deliberate. She left them *disturbed*, but not *directed*.

Though the council had no obligation to inform her, a woman and one not of their tribe, of their decision, after an hour someone came to report to her that the *jirga* had decided not to take the girl, that they would accept just a monetary compensation instead.

In the end, no one had lost, but all parties had gained. What had happened during those deliberations? We can deduce at least this: Samar's use of restraint, of avoiding judgment when sharing her thoughts with the court, helped her words to be heard. She began the encounter by showing her respect and even submitting to their culture, honoring their role as leaders. She offered two new pieces of information that were endorsed by respected leaders: that the practice was not actually condoned by Islam and that the tradition used to be carried out differently. She left room for them to arrive at their own conclusions after debating among themselves.

Samar's approach honored their onlyness. If she had conveyed her goal of realizing her own ideas as being more valuable than the *jirga*'s,* it would have been framed as a power-*over*-others model rather than the tougher but more enduring power-*with*-others model.

"Talking with this father after he learns his daughter will not be given as compensation was one of the most memorable experiences of my life," Samar said. "He just couldn't believe it. He didn't know who to thank because he could never think that I could have stopped [the *swara*]. Even I couldn't believe that."

There are no guaranteed methods to achieve a win-win in any conflict.

But what is almost certain is that when we believe the worst of others, they will often show us that side of themselves. Likewise, if you try to browbeat them, they're more prone to digging in their heels than shifting their thinking. In reality, you will never be able to convince everyone; there will always be resistors.[†] But you don't have to. It's easy to believe that you need 50 percent or 60 percent of a group to support you to create lasting change, but the latest research shows that the pivot point might begin with as little as 10 percent committed to push forward.[‡] You simply need to find those people who are commit-

---

[*]   This idea was inspired by Adam Kahane, who wrote in his book *Power and Love* (San Francisco: Berrett-Koehler, 2010) that "my drive to realize myself slips easily into valuing my self-realization above yours."

[†]   Cornell professor Shaul Oreg found there are four kinds of change resistors: (a) routine seekers who agree with statements like "I'd rather be bored than surprised"; (b) people who have strong negative emotions at the prospect of doing, or dealing with, new challenges; (c) short-term thinkers who agree with statements like "When someone presses me to change, I tend to resist even if I think the change may ultimately benefit me"; and (d) people prone to cognitive rigidity, who believe that "Once I come to a conclusion, I'm not likely to change my mind." See Shaul Oreg, "Resistance to Change: Developing an Individual Differences Measure," *Journal of Applied Psychology* 88, no. 4 (2003): 680–93, http://pluto.huji.ac.il/~oreg/files/jap_2003.pdf.

[‡]   Research done in 2015 by three professors in analytical fields (one in computer science, one in physics, and one in math) says that the prevailing majority opinion in a population can be reversed by a small fraction of committed members; specifically, at 10 percent. J. Xie et al., "Social Consensus Through the Influence of Committed Minorities," *Physical Review* 84, no. 1 (July 2011), https://pdfs.semanticscholar.org/21ce/52e518edef55a4eb05edb 19286132c5eb1a6.pdf. This number is useful but also seems too decisive because it doesn't

ted to a common purpose, the specific -*ness* that connects people together in onlyness.

CURIOSITY BLOWS AIR ON SPARKS OF CONNECTION

Respecting others who are fundamentally different from you is not easy, but when you begin with curiosity, you will have a useful tool to identify things you have in common and establish a starting point for an "us" to emerge. In Samar's case, she managed to find commonalities with the Pashtun tribal leaders: They all had a shared love of culture, deeply honored their religion and its practices, and wanted justice, even if they had different ways of achieving it.

How do you engage in a conversation as equals, with no "right" answer, but create the conditions necessary to foster an "us"? People often fear they will lose their own perspective if they abandon their opinions, but all you really have to give up is imposing your perspective on others. Establishing common ground involves making the following adaptations:

Seeking a Truth	⟶	Seek Our Solution
Making Accusations	⟶	Learning Perspectives
Creating Blame	⟶	Creating New Understanding

It is curiosity, and not contempt, that lets you create the drawbridge to unite the parties enough to find common ground. Once you establish common ground, you might be able to distinguish a common purpose, an us that is inclusive of all parties' interests. It's through a common purpose that people start to see themselves and others as one, which leads to cooperation. That cooperation leads to a series of actions, and those actions lead to systemic change.

---

take into account counterissues. For example, on some issues, such as gun control in the United States, "support" doesn't matter, as there is high support for background checks, yet no related social actions such as new laws.

TALK AMONGST YOURSELVES

Today, more than ten years after her first documentary started her filmmaking career, Samar hasn't stopped, despite pressure to do so. In 2009, she received death threats after being accused by a leading journalist in Pakistan of being a CIA operative, on the American payroll, because she highlighted a video about the public flogging of a girl in one particular village—watched by hundreds of its residents—and condemned the act on television.[17] After she was labeled an "Islam *dushman*" (enemy) by the local *jirga*, she was tempted to go into hiding, but decided against it and continued her campaign to engage people in her films and get them to consider their implications.

To make certain that the law forbidding *swara* was implemented, she took the cause to Pakistanis of all backgrounds—students, members of law enforcement, and even truck drivers—by screening her films.

"I'll show a film of a different village than the one I'm at, of a girl talking about why she doesn't want to have a forced marriage, or of a mother talking of how *swara* hurts a community. I keep the screenings in the local dialect. So if it's a Pashtun village, the film is in Pashto." In every group she addresses, Samar enables people to listen by inviting a conversation. Listening may appear to be a passive act, yet it's anything but. Research has shown that listening is the single biggest factor in accounting for successful leadership, besides basic competence.*

Conversations work because they are the place where all minds can change.

Franklin and Samar played an important role in redirecting and

---

* Forty percent of the variance in what makes someone a good leader is tied to listening. That's the single biggest variance other than basic competence. Other research says that the average person listens at only 25 percent effectiveness. In a study of eight thousand people, most believed they were more effective listeners than they are. See Robert Kramer, "Leading by Listening: An Empirical Test of Carl Rogers's Theory of Human Relationships Using Interpersonal Assessments of Leaders by Followers" (PhD dissertation, George Washington University, February 1997), abstract on The Leadership Challenge, accessed February 21, 2017, http://www.leadershipchallenge.com/Research-section-Others -Research-Detail/abstract-kramer---leading-by-listening.aspx.

reframing the conversation that was already in place. They were deeply respectful of their audiences and their viewpoints but also opposed the notion that the existing answers were the only answers. I call this particular role "the loyal oppositionist"*—a phrase I introduce to my colleagues and friends when I want them not only to challenge my thinking on a particular subject but to *also* help me to come up with better answers and ideas. The important part is their productive approach as they do the challenging; their job is to encourage the idea to get better. In that way, the approach isn't to make someone feel wrong or less than for not already being clear but to want better for the idea. As gratifying as it is to have friends simply tell you, "Wow, what you're doing is wonderful," few of us need more "yes" men and women in our lives. What we *do* need are people who play a multiplicity of roles: to cheer us on, challenge us to do better, *and* even admonish us if we're not living up to our ideals.

The battle to shape the future isn't something done to another, asking another to capitulate to our side. Instead, it is carved out from new questions and seeking our common purpose, something we find as we wrestle with ideas, together.

---

* The term "loyal opposition" has political roots; see *Merriam-Webster's Dictionary Online*, s.v. "loyal opposition," https://www.merriam-webster.com/dictionary/loyal%20opposition, accessed February 21, 2017. It reflects how two or more parties can have very different ideas but still want the best for the country.

CHAPTER 6

Without Trust, You Don't Scale

Hell is a place where nothing connects with nothing.
—DANTE

TED FELLOWS: BELONGING COMES AT A COST

Even though Tom Rielly didn't want to, he fired Eddie Huang.

When Eddie violated the rules of the TED Fellows program by sneaking out on day four of the seven-day-long TED conference, Tom got e-mails and texts from the other nineteen fellows notifying him of Eddie's absence. One of the few agreements of the program is that participants remain for the entire week.

"What's up?" Tom asked when he finally got the fugitive on the phone, as Eddie was driving to a PR gig in Los Angeles. Hoping for a rationale that would explain the prior twenty-eight hours, which Eddie had recorded on Instagram, Tom wanted to know why he had gone AWOL to attend a Lakers game and have dinner with friends. Though Tom urged him to return, Eddie said he had a few more things to do. This was too much for Tom, who had to let him go on the phone.

Eddie subsequently used a radio interview he had been on his way to as an opportunity to talk about the experience and what it implied about the famous TED conference.[1]

"It's a cult," Eddie said. "I just went through a whole week of people telling me what to do and where to be. It was horrible . . . Every day they have thirteen hours of ****ing activities mapped out for you." Eddie Huang, a restaurateur and author of the memoir *Fresh Off the Boat*, clearly hadn't felt a bond to the group he had willingly joined, a group he was now calling a "Scientology summer camp."

We've all been a part of a group in which people come and go as they please. They treat the group like many of us treat our gym memberships: something that's always there for us, for our convenience. They "belong," but it is entirely on their own terms. If we're lucky, we've also been part of groups where we are missed when we're not present, whose members we share dreams with, and worry about, and congratulate on their successes. The dynamic of the latter type of group doesn't happen automatically but requires an investment of something: ourselves.

Did Tom have any regrets about kicking Eddie out? Years have gone by since this 2013 event. Tom sighs, looks away, and takes a moment to reflect before he answers. I wonder if he's thinking about the more recent news that Eddie also got into a public fight with the people who turned his book into a television show; the one which the *New York Times Magazine* wrote up as "Eddie Huang Against the World."[2] But Tom doesn't mention that as he answers. "The decision was the right one," Tom explains, "because Eddie wasn't a collaborator." But, he adds, "If I did it all over again, I would have been perhaps less emotional about it."

The seeds of the TED Fellows program were planted in 2007, when a hundred young innovators were invited as guests to attend the conference TEDAfrica. One of the participants was nineteen-year-old William Kamkwamba, who would come to change Tom's life.[3] When he was just fourteen, William had taught himself how to build a windmill from objects he found in his local junkyard, working from rough plans he found in a library book. He hadn't been able to go to school at the time, because his country, Malawi, was experiencing a famine and his farming family didn't have the income to keep him there. Tom and the entire TED audience were deeply moved by William's story, and so he and the other conference organizers started to think of ways to include people like William—fresh new voices achieving amazing things—as a regular part of TED.

TED, which is an acronym for "technology, entertainment, and design," has grown to include any "ideas worth spreading." It celebrated its thirtieth anniversary in 2015, but when Chris Anderson, a successful Oxford-educated media mogul, acquired control of it in 2002, it was simply a conference for an elite crowd, a forum for a few hundred mostly rich attendees from the United States. Chris wanted to keep the best of what worked while also infusing the gathering with new energy and fresh perspectives. He added a global lens by sponsoring onetime events like TEDAfrica, and TEDIndia, and the now yearly TEDGlobal. He also authorized the sharing of the presentations delivered at the conferences (TED talks) online, free to anyone, a turning point that would dramatically extend the organization's impact. Chris and his team also initiated the TEDx construct, which would allow anyone to independently organize a local conference to share valuable ideas with local communities. More than ten thousand such events have been held at the time of this writing.

Inspired by the talks at TEDAfrica, Tom convinced Chris of the value of building a full-fledged fellows program. As director of community, he had been responsible for approving every application for conference attendance, so he was ideally situated to judge what TED needed. He had already gained a reputation for community building in his role as founder and chairman of PlanetOut Corporation, then the largest online resource for the LGBT community. Tom, having experienced the difficulty of telling his own parents that he was gay, always managed to create spaces for people to be welcomed. Karen Wickre, a former Google and Twitter executive who has known Tom for decades, believes that his onlyness "comes from his capacious brain and heart. They encompass his deep curiosity about the world and everyone he meets. Rare is the person who doesn't think of him as a friend from the first encounter. But [Tom's talents lie in] also knowing ways to create some context for doing more [together] than each could do on their own."

Tom was ready and willing to translate his TED Fellows vision into reality. But what it would become, and how that alchemy would happen, would, using André's word, unfold.

THE BONDS THAT TIE US TOGETHER

If the people who move ideas into reality don't know how to be an "us," no idea stands a chance of succeeding. The "power of us," then, is not the sheer number of members in any group but the bonds between them, such as curiosity, vulnerability, the ability to handle a conflict. To move an idea into reality, everyone involved with it needs to know how to be *curious* enough to discover the right problem to solve. They need to *listen* to one another as options are explored, and to be *vulnerable* enough to accept help from one another. Also, they need to *tussle* together on tough decisions so that, ultimately, they can *lean on one another* as they prepare to move into action. Without those bonds, everyone is effectively running solo, staying unconnected from those who can help them be better, do better, and make bigger outcomes.

So, how do you get bonds to happen?

WHY ME?

"Oh, my God, how did I get into this room?" Negin Farsad, a comedic performer of Iranian descent based in New York City and the director of the film *The Muslims Are Coming!* remembers thinking about being recruited as a 2013 TED Fellow. She describes her experience much like other TED Fellows: "Every one of us wonders what we're doing there, how we got there. Ninety-nine percent of the fellows just don't have the financial means to go to TED on their own—the ticket alone starts at seventy-five-hundred dollars. We're really capable in some relatively narrow niche, but we don't have the pile of credentials or monies to be there [independently of the fellowship]." Another fellow, network scientist Eric Berlow, echoed Negin's sentiments, as he, too, "wondered why I was even being considered."

To find these kinds of recruits, Tom and his colleagues begin by seeking varied people across disciplines and then sorting them with some form of these questions: "What is it you don't want others to know about you? What is that thing about you that you hate? What is it you suck at?"

These seem similar to the "what are your weaknesses" questions typically asked in a job interview. The responses, such as *I work too hard*, are rarely revealing, and the exchange almost becomes a game in how not to say *anything* of interest. Anything too weird or too wild might get rejected from consideration.

For his part, Tom wants to learn not "how little you reveal, but how well you reveal yourself. I'm listening for a lot of things," he explains. "How self-aware you are, how humble you are, how vulnerable and real you are willing to be." Because, he argues, "if you give the fake-out corporate-type answer, you're probably not very open—and TED Fellows needs open."

Tom sees "open" as permeability to the world as it's changing.

Most job recruiters want to ascertain whether you've done enough work and put in the ten-thousand-hour equivalent of what it takes to do a predefined job. Tom is certainly interested in a commitment to and excellence in a given field, but he's clearly looking beyond smarts. Smart measures that you know *enough*. But to be open is to want to learn from others. "Because as good as you [individually] are, you can't get any better without being open to others," Tom explains. "And the only way you can be open is to fully embrace that which is *you* . . . your fullest, quirkiest you."

ICEBREAKING

After the twenty fellows are accepted into the program each January, they attend a three-day retreat the following month to meet their colleagues. Instead of being introduced to one another through professional profiles and accomplishments, Tom and his deputy director, Logan McClure,* set a warmer relational context.

Tom and Logan introduce the fellows to one another by telling stories of each of them, always including some quirky anecdote or explanation about why they had been picked. (Logan joined Tom very early in the TED Fellows program; she had earlier run a nonprofit in college to build an all-girls school in Tanzania. The fellows refer to Logan as their "mama," even though she is only in her twenties.) "We want to help each fellow see that which is distinctly true about the person being introduced—not just the obvious points of difference," Tom says of the process, adding that this method is better than self-introductions. "Knowing people can read bios of accomplishments, we focus on what the essence of the person is, because usually they can't or won't name it." Each of the fellows is then encouraged to add anything about themselves that they wish.

While you can't automatically create a community, you can create the conditions for one, which is what Tom is doing. To get to know one another, you have to see one another. And to see, you have to allow yourself to be seen, beyond the façade.

Although not all of us have the luxury of a three-day off-site to make acquaintances, we can, after an introduction to a new person, make a connection with *her*, and not the particular role she is playing. Introduce yourself by expressing your passions and interests, not reciting your job title or employment history. Open yourself to other people to allow them to see you, even if, and especially if, you're quirky.

---

* Logan worked with Tom from 2008 to 2013, starting as his program manager.

This creates the opportunity for any two individuals, as well as larger groups, to build a bond based on something of meaning.

"Tom sets the context with such love and respect so that people *can* see each other," says Sunny Bates, a twenty-year-and-counting "TEDster" and one of twenty select people named as part of the TED Brain Trust. She's participated in each of the TED Fellows programs as part godmother and part super-connector. As a former executive head-hunter, she has a long history of knowing how to find talent. "Each person chosen into the TED Fellows program is at the cutting edge of their field," she explains, "but almost always an interloper or weirdo outsider in that professional field. They feel 'the other' in their work, typically alone, an outcast. No one or very few get them. If there is one word to describe the Fellows program, it's love."

Sunny's observation uses a word not often used in professional re-lationships. But love is the other-acknowledging, other-respecting, other-helping drive that reunites the separated,[4] as Adam Kahane wrote in his book *Power and Love.* Love is not the exclusive property of couples; it can also be an asset in groups, like ones organized in only-ness. Love as a private asset—between two people—shows up as pas-sion, excitement, and even commitment. Love can be an equally strong asset in a community, among many people who belong to a group. Love as a community asset manifests as meaningful bonds, caring and respect, and, ultimately, the trust needed to count on one another.

Without Tom's intro, the fellows' stories suggest, there would likely be a tendency to either strut or apologize—strut to justify why they are there or apologize for not living up to expectations. Whenever in-dividuals are thrown into a situation without a clear understanding of who we are to one another, the first human response is to define a hi-erarchy, a pecking order.* This is what the strutting or the apologizing

---

* Jeff Pfeffer has offered key insights on why teams revert to hierarchical models; see Eilene Zimmerman, "Jeffrey Pfeffer: Do Workplace Hierarchies Still Matter?" *Insights by Stanford Business*, March 24, 2014, https://www.gsb.stanford.edu/insights/jeffrey -pfeffer-do-workplace-hierarchies-still-matter.

is about. The more turbulent the situation, the more strutting typically happens as people jockey to create a sense of place in the group. Unless the conditions are set up correctly to allow for better norms, pecking order norms are the default.

But when the better norms have been established, when we love someone *just as they are*, then they know that they are being appreciated for their own selves. Those who feel accepted can step outside of themselves to engage openly and freely, to be curious about other people and not feel the need to prove themselves. But if they feel raw, or judged, or unworthy, they're far more likely to withdraw into themselves.

Ultimately, what makes a collection of people "yours" is how you feel about it. The key quality that makes a group or tribe or community yours is that for you it becomes a place that you can shape and be shaped by. The bonds—of caring, of trust, of love, and, of course, of purpose—make your community a sanctuary where you can freely admit you don't have it all together and where you can be willing to ask for or give help. It is precisely for this reason that groups like families, churches, causes, or high-performance teams are demanding: In addition to our time and our brains, they require us to give of ourselves, and be open to be given to.

CHOOSE HOW YOU WILL BE SHAPED

Although Tom plays a major role in making the TED Fellows program into a community, his original vision didn't take final shape until his own community helped it do so, refining his vision for it.

One member of the current Fellows community is someone Tom knew from the broader TEDster tribe—Ruth Ann Harnisch, a journalist-turned-philanthropist and founder of the Harnisch Foundation. When Tom and Logan first met with Ruth Ann, in 2008, the TED Fellows program was, as yet, unfunded, and Ruth Ann became the first dollar in.

As well as funding, Ruth Ann brought to the project moxie and ideas. She had experience with several different prominent fellowship programs and knew which of them had made a significant dent. She had seen how pairing people with one-on-one mentors or coaches made a huge difference in overall effectiveness and pointed to work she was doing with the International Women's Forum as a good example. Ruth Ann decided that if she was going to support this new Fellows program at TED, she would take a role in maximizing its growth and development, and so she proposed: "Let's build in a coaching program for the fellows. There are at least fifteen nameable differences that a coach can make," Ruth Ann explained,* and proceeded to argue why the fellows deserved and needed coaches to support them to foster more open relationships, more growth, and more leaning on one another, a vision that matched Tom's own goals.

Though he didn't originally get it, Tom now views coaching as a crucial part of the Fellows program. "Lots of fellowships don't delve into the psychology and emotions of their fellows at all. They might think it's impermissible, it's over the top, or it's beyond the pale, or whatever. The truth is the fellows are human beings. And every human being—high-performing and just regular people, whatever—have stuff, have baggage, and we all need support and space to do our best work." Today, 150 coaches and mentors are part of the extended TED Fellows team in an adjacent program called SupporTED, which has enabled the Fellows program to scale up coaching so it doesn't rely on just a few people's energy being spread thinner and thinner.

Coaching, Tom now contends, is "like an exoskeleton of belief," a space for people to step into themselves, a space to grow, to become more who they want to be. I call the tribe that provides this for me my "tent-pole holders": Unless there's a pole holding up the fabric, the structure collapses and leaves no space inside. When I'm all up on

---

* When Ruth Ann became a coach, she had to understand the Fifteen Proficiencies, on which she was tested when she received her coaching certification from the International Association of Coaching. All of them come from Thomas Leonard's models. More information can be found on the website of the company Leonard founded, Coachville: http://www.coachville.com/.

a project, or feeling as if I'm shrinking from the weight of a task, with no visibility for where it's all going, I'll ask a friend to hold the tent pole for me, providing the emotional support and psychological space for me to find my way into a new approach or solution to a problem.

While Tom accepted Ruth Ann's idea to get the initial funds he needed—$50,000 a year for three years—to begin the TED Fellows program, it's just as telling that he didn't agree to every dollar that came with conditions. Google came to him with a tempting offer of support, saying, "We'll give you money to fund a certain amount of social entrepreneurs in the group." Tom's response was very clear: "No, we are not a social entrepreneurship fellowship. We are not a technology fellowship. This is a fellowship that is extremely cross-disciplinary, that ranges from activists to inventors. That's what makes it special. We're going to protect that really hard." Both Google and Ruth Ann had funding *and* ideas to offer, but one enhanced the core idea of the program, while the other took it in a new direction.

PICK AND CHOOSE

Does any group have to let you do whatever you want, and if they don't, they're limiting your onlyness?

No.

Just like there is no such thing as a perfect mate—there's only a perfect mate for you—no group is perfect; it's only a perfect one for *you*. We each have to choose which is a suitable fit for our own needs and goals. A group will not become a community for you—the "you" doesn't morph into "yours"—until you agree to the core aspects of its values, goals, and conventions, you won't belong.

The key is in the agreement: the implied handshake between your purpose and the group's shared purpose, between what you give and what you get. In the TED Fellows example, Eddie Huang had agreed to the clearly stated norms and conventions in exchange for certain benefits, yet he chose not to abide by them. One could misinterpret what the Fellows program did in response as "denying onlyness." But, in fact, that would not recognize the big difference between *allowing self-expression* and *denying selfishness*. The other nineteen fellows were also expressing their onlyness, but their common interest was also in supporting one another. Eddie's apparent self-interest conflicted with the shared purpose of the other fellows.

When a bond links people together through mutual understanding, reliance, and support, everyone benefits, individually and as a whole. The solid strength of the whole gives strength to individual members. It is these trusted bonds that make it possible for individuals to come together, to make a dent, without sacrificing group members' passionate ideas.

ROLL TWENTY DEEP

Let's review the five lessons we've learned from the TED Fellows experience that promote tighter bonds: (1) Design a process to pick people who are open to one another. (2) Enable people to see one another, with all their strengths and flaws. (3) Be open to changing yourself and letting others shape your ideas. (4) Yet, know what to say no to. (5) Make decisions about what behavior compromises the community.

You might use different techniques to introduce people to one another or foster a sense of openness, but recognize that this is the work that must be done.

The individual who enables the healthy dynamics of community is essentially an alchemist. She pays attention, shifts priorities, blends, and creates a context where everyone can be more than he or she is alone. While none of us can *make* any set of people bond, we can *manage* the context and dynamics to encourage bonding.

What did the TED Fellows ultimately accomplish because of their strong bonds?

- An underwater-robot inventor collaborated with an African technologist to leverage a political-crisis reporting tool and build a first-ever open-source community of scientific explorers.
- An ophthalmologist met an ocean conservation scientist, who helped him create a marine alphabet for the blind.
- A magazine editor/designer met a computational biologist, and together they created an infographic about rare genetic diseases.
- A Jordanian social entrepreneur and philanthropist collaborated with a strategist/advisor to help open the door for the airlift evacuation of tens of thousands of injured civilians during the 2011 Libyan revolution.
- A tissue engineer and an architect formed a biohacking hub to enable the future of building with biology.

Before the existence of TED Fellows, so many of these individuals were laboring in isolation. Now they have a community. We've already talked about the value and importance of not being the "only one." If you feel alone, you will not pursue an innovative project as vigorously, because the need to belong is greater than the desire to create. The support of a trusted community enables people to dream bigger dreams—and ask for help, accept help, push one another, and hold one another accountable.

Negin describes how walking into the TED conference now as a senior fellow has given her the confidence to face the billionaires and news makers taking part in it as equals: *Yeah, you might be rich and powerful, but I've got my [TED Fellows] posse here, and they have my back. And with them, I roll like twenty deep.*

TAKE A DIFFERENT PATH HOME

It's easy to be confused about what facts to pay attention to, or whom to count on, in a world that is so full of noise. When ideas can now scale quickly and powerfully through the strength of a network of people, it's increasingly important to know whom to lean on, what to believe, and whom to trust.

Our ability as humans to organize, specialize, and cooperate is especially valuable in life-or-death situations. When, for example, in prehistoric times a large group went hunting and two of its members saw a lion in the tall grass, it made sense for them to set aside any doubts about their companions, trust one another, and take another route back to camp.* The decision might have been the wrong one, perhaps influenced by bad information, but on balance, in such life-or-death situations, trust served our ancestors well.

Something similar is taking place with today's online tribes. We're all sharing information and pooling insights, a fundamental tool that helps us humans thrive. When many people can draw on the knowledge and experience of the collective group, we all gain from each other's talents, ideas, and specialization—just as Franklin Leonard demonstrated with the Black List.

But we can also be fooled and coerced by drawing on the wrong people, who shouldn't be trusted.

In the past ten years, childhood immunization, which had long been an accepted treatment that conquered dozens of diseases, has become a debatable topic—not because vaccines have been scientifically proven to be harmful or ineffective but because of the size of the faction of anti-vaxxers. Those skeptics ask: What's contained in those vials of vaccine? Is vaccination really in the best interest of the global

---

* This "lion in the grass analogy" was inspired by Daniel Gardner's book *The Science of Fear: How the Culture of Fear Manipulates Your Brain* (New York: Dutton, 2008). He pointed out that in tribal days, fear served a lifesaving purpose.

population, or is it that pharmaceutical companies are simply chasing the almighty dollar, with the government's support?

The seed that sprouted the anti-vaccination movement is a five-page research paper published in the British medical journal *The Lancet* in 1998 by a man named Andrew Wakefield and twelve co-authors. The study falsely claimed that twelve anonymous children began to show symptoms of autism two weeks after receiving the measles, mumps, and rubella (MMR) vaccination. Later shown to have been a complete fabrication, it was formally retracted by *The Lancet* in 2010. The truth about the motivations for Wakefield's "research," and the elaborate PR campaign he launched that spread the vaccines-cause-autism myth, was revealed by an investigation led by reporter Brian Deer.[5] Deer discovered that Wakefield had been hired by a lawyer to discredit the MMR vaccine years before. The not-randomly-selected children who supposedly developed autism were never even given the vaccine. The widespread panic over vaccines and the resulting reduction in vaccine administration worldwide made Wakefield and the lawyer millions of dollars.

During the height of the anti-vaccination movement, celebrities who had autistic children lent credibility to the fallacious claims, and media outlets jumped on the story, blowing it out of proportion. It was a perfect example of a contemporary lion in the grass—but, in this case, the threat was not real. Even though it's been proven false by one of the largest studies ever (one that included 1.25 million children),[6] the story lives on—online and in the minds of scared parents who have begun opting their children out of receiving vaccinations for life-threatening illnesses.

This account ultimately concerns more than just the issue of the safety of vaccines, for it raises a critical question that we face constantly in the media: Whom do we choose to believe?

Do you believe everyone who is challenging conventional thinking? After all, social progress takes place as the result of questioning truths that others take for granted. When we observe and act on something based on our onlyness, we often create new solutions to old problems, or entirely new ideas never conceived of before. Or do you trust those in authority, who have established their expertise in a given area? Doesn't expertise matter? What facts count? Is every example of self-expression credible? Is every opinion out there equal in value? Ultimately, how *do* you decide what opinion to listen to?

ERODED TRUST

The question of whom to trust is not as easy to answer as it once was. Trust in many of our established institutions—churches, businesses, governments, the military—has decreased. According to Pew, trust in government in the United States is at a near fifty-year low.[7]

Have conditions really changed so dramatically, or have we all just grown more cynical?

It's likely that both are true. As we collectively lose trust in all forms of authority, we begin to doubt every source of "official" news. It's understandable that people have turned to Internet chat rooms, which are often home to conspiracy theories of various sorts, to learn "the truth." This has become a problem for all of us, and unless we learn how to filter out the truth from convenient fallacies, we risk losing out on all the benefits that groups provide through the platform they provide for cooperation.

How do we make responsible decisions about where to place our trust where ideas are concerned? Here are some suggestions.

If the question is fundamentally subjective—for example, "Does Uncle Clay make better berry pies than Aunt Margaret?"—then your opinion typically counts as much as the next person's, even if that person is a doctor with more degrees than you. Your vote matters in this situation because the issue at hand is a matter of preferences. You're likely to face this sort of decision every day, such as when you have to choose which coffeemaker to buy. Your selection will depend on how you plan to use the device. You might prefer using preground beans, which will limit the number of options you'll have. But if you're a coffee connoisseur, preferring to consider factors like the temperature of the water for brewing, the exact provenance of beans used, and which grind is appropriate, you'll have a different set of priorities.

If the topic is dogmatic in nature—as in, "Is there a heaven or a hell?" or even "Is there a God?"—the question is not one dependent on expertise or facts. There is no "right" answer other than the one each of us discovers for him- or herself. That's not to say other people don't matter in this search—they do. Our friends can give us moral support and help us talk through our ideas. But at our most vulnerable moments, we can also be falsely directed. Online militant groups have recruited young people who feel lost in life and encouraged them to engage in destructive actions as a way to make their lives matter.[8] When we listen to someone else for advice about what can give our lives meaning, we are literally handing them authority to define who we are and what we become.

If it is a question about the future—"When will cars drive themselves?" or "What's the market opportunity for our product in the next ten years in Eritrea?"—then you'll want to draw on the expertise of others and to combine and interpret different data, insights, perspectives, and observations to make an educated prediction. Someone who has lived in Eritrea might have as valuable a local perspective as the person who has done ten product launches in a similar country. Future-facing issues are almost, by definition, a moving target shaped as things evolve. So be aware that whatever advice you do get is a matter of forecast, not facts.

For objective questions for which the scientific method applies—on questions like, "Is climate change taking place?" or "Are vaccines harmful?"—actual hard facts data can serve as a foundation, with different people taking their own individual stands on the topic at hand. In the case of global warming, for example, 97 percent of climate scientists agree that global warming is actually happening, and that it's a man-made disaster.[9] Even though there is a great deal of noise circulating that suggests the issue is a debate, it's actually not. The misinformation surrounding it is only a subterfuge, likely advanced by those who want to preserve the status quo. Figuring out who has what to gain around these debates can help you navigate to the more credible sources.

Trust is a powerful human trait, and relationships and bonds are key assets, so though they have been abused and eroded, we shouldn't abandon them. Ultimately, we each have to develop better skills to discern for ourselves whom to trust, and for what. It's hard to lean on other people; doing so makes us vulnerable, and we can be duped. Because none of us likes feeling foolish, it can seem safer and somehow wiser to rely on our own counsel—or at least on that of people who are most like us. But to do so in all cases would be imprudent, because it would mean we weren't taking in new information that would help us do better, together.

PATIENTS LIKE ME: CHANGING THE EQUATION

"So you know," said Dr. Brown to Stephen Heywood, "the, uh, diagnosis, unfortunately—that all of [these tests] point toward—is ALS."

ALS—amyotrophic lateral sclerosis, or Lou Gehrig's disease—is rare. It's also 100 percent fatal. On January 16, 1999, thirty-year-old Stephen was effectively told he was going to die, probably within five years. Over lunch that bleak winter day in Boston, near Massachusetts General Hospital, Stephen sat stunned with his mother and older brother Jamie, who could not keep himself from crying. He couldn't accept the news of Stephen's death sentence and vowed to save his brother's life.

Jamie is a mechanical engineer with dyslexia, whose love for his family was everything. For five years, he pursued a scientific breakthrough, entrepreneurially raising nearly $50 million to develop bleeding-edge gene therapy and stem-cell research. Despite profiles in the *New Yorker,*[10] the *Wall Street Journal*, and the *New York Times Magazine*, on *60 Minutes*, and even in a book written by a Pulitzer Prize–winning science writer (*His Brother's Keeper*), his valiant efforts were ultimately unable to make a difference for Stephen. Jamie's determination cost him dearly: His marriage dissolved, and he burned through all the money.

In the process of chasing many dead ends, however, one of his efforts turned out to be enduring: an online network to help ALS patients. This early formation of a community of patients proved to be the most effective strategy to make a dent through the *power of us.*

The community's proposition to its membership was essentially this: "Give us your personal health information for free, and we will sell it at a tidy profit, to advance research." Does that sound attractive, or even compelling? It turned out that many people would agree to those terms *if*—and this was a big *if*—they trusted those involved. Jamie's dent was to use community to change the equation for curing previously incurable diseases—but first he had to enable that trust.

Jamie and his brother Ben created the social networking site to save Stephen, PatientsLikeMe, back in 2004. Today, PatientsLikeMe is a for-profit organization with the largest and most active patient community on the Web—500,000 contributors reporting on their real-world experiences of their particular illnesses. From its original singular focus on ALS, it has since expanded to include 2,500 conditions. What all its members have in common are diseases for which medicine doesn't have answers today. Some of these are "orphan diseases," which are defined as conditions that affect only 1 out of 2,000 people in Europe, or fewer than 200,000 people in total nationwide in the United States. Orphan diseases are considered too expensive to be "worth solving." The pharmaceutical and biotech industries largely ignore them, because even if they do find a treatment, the size of the market means there's little financial incentive as compared to "bigger" diseases to get a return on their research and development.

NEVER ALONE

"At first, I came to [PatientsLikeMe] because I could learn so much for myself," says Ed Sikov, who has Parkinson's. Even though Ed now lives a healthy life, as he's able to manage his symptoms through medication, he still visits PatientsLikeMe every day, his first stop after checking his e-mail, because, as Ed explains, it can "help the next guy." His hope is that by giving his data to science, he can advance progress to find a cure: "Then, I am more than a victim of a disease, I am a problem solver of the disease." Ed shares personal data—what medicines he's taking, their side effects, if he's depressed, if he's gaining weight, and even his sexual activity.

"I used to feel alone before PatientsLikeMe," Ed says. He values the site's providing an aggregation of many people's data. "When I share information, I help others. It might become clear that one treatment works better for people my age, or that a simple shift of when I take a drug changes how I respond. Sharing my own data lets all of us have more insight.

"I could be suffering for twenty years, but if I could prevent the next person from suffering for twenty years, then my suffering isn't for naught." Sharing gives any suffering he does experience more meaning by making it more "worth it."

Ed's personal "only" that shapes his commitment to the community is his belief in the power of patients having a voice. When he was about nine or ten, his primary caretaker was an aunt. He remembers accompanying her to see her physician when she was sick. "The doctor was one of those sorts of old-school doctors that acted as if they knew everything and had the attitude of, 'Just leave it to me and I'll take care of it,'" Ed recalls. His aunt kept describing her problems, but the doctor barely listened before he dismissed her; she ended up dying of a related condition shortly thereafter. Ed also recalls vividly how his aunt felt, and—most importantly, perhaps—what she knew about herself.

"Maybe she wouldn't have died," he acknowledges, "but that's not the point. If the doctor had listened to her, at least she would have felt heard, she would have felt seen, and for her and for most of us, that matters. It changes the very quality of our life. It changes whether or not we're being witnessed in the world or whether we matter to someone else."

A DIFFERENT NOTION OF SCALE

"Social changes—big changes—happen because ordinary people are willing to work together, and those people only contribute and participate because they want to," says Dr. Paul Wicks, now vice president of innovation at PatientsLikeMe.

Paul joined as a community manager; he was Jamie Heywood's sixth hire. Trained as a neuropsychologist, Paul met Jamie after he had been moderating an ALS online forum for five years in London. At the time, PatientsLikeMe had only fifty ALS patients, which nevertheless was an order of magnitude larger than any other group Paul was aware of. "It used to take labs weeks or months, and sometimes years, to find a sample size of a hundred patients to do relevant research," Paul says. "But now, I can push a button and survey two hundred ALS patients and get results in two weeks."

One of the greatest services provided by PatientsLikeMe is its ability to scale.

Scalability has traditionally been concerned with the ability to cope and perform well when faced with an increasing or expanding workload. A chip manufacturing firm, for example, scales well if they can keep up with increased demand while keeping quality up and maintaining or improving profits. In other words, *uniformity* has traditionally been a central feature of successful scaling. PatientsLikeMe involves a different notion of scale. It's the scaling of onlyness: a high degree of *variety*, but unity in a shared idea, so that together, the group creates the *mass* necessary to bring about change.

We've seen the same type of scale in Alex Hillman's quest to make Philadelphia welcoming. In the story of Rachel Sklar and Glynnis MacNicol, the scale of many changed the ratio of women in tech. The scale of the many contributions to the Black List helped Hollywood achieve its deeper aspirations. Samar's efforts against *swara* brought many together to align justice to a higher set of values. In the story of Tom Rielly and the TED Fellows, it created the bonds that enabled creative collaborations to develop. Bound by shared purpose, these united groups of networked people,* aligned around an idea, have accomplished what once only large organizations could do.

---

* "Networking" is different than "networked" or "networks." An example: Another author was recently introduced to me. After doing a few minutes of homework (watching his recent TED talk, scanning the outline of his book), we exchanged notes. After expressing to him how we seemed to have two things in common (economic prosperity and the sense that tech needed values-based design), his disappointing response was to ask if his recent book could be plugged. I was focused on purpose-based networks. He was doing transactional networking. Transactions can certainly be fruitful; one helps the other with an implied IOU for a parallel transaction. But that is different than belonging to the same network, which has shared goals, or common purpose, in mind.

The power of numbers means that PatientsLikeMe makes research for rare diseases more cost-effective, as Paul and his team develop protocols so that patients can answer questions easily and without jargon, to turn anecdotes into data. Previously immeasurable "soft" health issues like nausea, mood, fatigue, and quality of life can now be quantified through questionnaires. A contrast-sensitivity test has been developed with Massachusetts Eye and Ear for people with Parkinson's to track the hallucinations that accompany mood disorders. One test designed with TED Fellow Max Little involves calling a phone number and talking into a recorder to assess the progression of Parkinson's through changes in the voice. One survey tool scans language in chat boards so that if someone comments, "No way would I ever do that again," the anecdotal account can be presented in a form that others can learn from. PatientsLikeMe publishes research papers that are peer reviewed to advance research. At the time of this writing, they've already published more than eighty such studies. Genentech, AstraZeneca, and more than twenty other firms are now working with PatientsLikeMe.

It used to be that a good doctor could make a significant difference in the treatment and outcome of a disease, while patients were, relatively speaking, not in a position to help much with their own conditions. As medicine has grown more sophisticated, patients now have greater access to their own health data. And now PatientsLikeMe allows them to aggregate their insights to solve things, together. It's a shift in power from experts being in charge to each person counting.

SOCIAL CONTRACT

PatientsLikeMe's website describes the organization as "a for-profit company (with a not-just-for-profit attitude)."

"Part of our belief is that nonprofits actually have a systemic problem," Jamie explains. He ran a nonprofit for ten years, so he has credibility on this topic. "The more you succeed in growing, the more of a problem you have, because then you need to raise monies to serve new people. The more successful you are, the more you succeed, the bigger gap you face between the work you need to do and your ability to serve that work. And, so, [one can] end up spending a disproportionate amount of time raising monies in small quantities to [be able to] do [the] work that matters.

"There's no clear black-and-white answer here of what path we should have taken to set up this firm. There [are] strengths, vanities, weaknesses, and errors in all the business models. There's no escaping that truth. To believe that nonprofit is ideal is a false high horse."

The issue Jamie raises is a question that many people involved in providing social services eventually have to ask themselves: Is it morally acceptable to make money when solving a problem?

"The key was for us to create an economic architecture so we could both grow fast and make a difference in this space, but also so we could be sustainable," Jamie explains. Scale gets you the momentum you need to make a dent; sustainability lets you do so over time. You need to design for both.

"One of the things we have on every page of the website is how we make money," Jamie says. *Every page.* We want the patients to know, because we believe in transparency. We tell our members what we do and do not do with their data. And then we sell data that engages partners in conversations about patient needs, to help them better understand the real-world medical value of their products or to accelerate the development of new solutions for new patients."

What is PatientsLikeMe enabling with their transparency? Trust.

As connected people can now gather to get things done, it requires us to be in relationships that are predicated on trust. Trust is not something that can be imposed; it's something both parties in a relationship have to come to on their own. There is an implicit formula—informed by your previous interactions and experiences—that you use to evaluate whether to put your faith in others. You'll typically ask yourself such questions as:

Will they do what they say they're going to?

Can they even honor what they say?

Are their intentions and mission clear?

Will the leader choose self-interest over that of the group's interest, so that all the time, energy, and passion one puts in will be for naught because of selfish behavior?

These questions formulate the mental math of the trust equation,* a way that each of us evaluates if we're willing to put energy into "us."

Without trust, we're just a disparate group of individuals who happen to be gathered and hanging out in the same general place; the power of *us* is diminished without it. Without trust in each other, we lack the fulcrum on which the many join forces to make a dent. Trust also enables you to do things more efficiently because you don't have to spend time preplanning the *can I* or *should I*, or weighing potential consequences. You just *do*. This is what allows the scale of onlyness: to go faster, together. The opposite is also true. The lack of trust results in obstacles like monitoring, endless negotiations, bickering, and ongoing disputes. These slow everything down and cause people to focus on their own interests, not the group's. Without trust, ideas are not shared, not built up, and certainly not made real.

Jing Zhou, of Rice University, studied the effect of trust on performance within organizations, yet her results are more broadly applicable. She found that trust is critical: "If my supervisor demonstrates empowering leadership, but I don't feel it's genuine, I'm not going to take the risk to be creative . . . As a psychological state, creative self-efficacy usually precedes the behavioral outcome that is creativity."[11]

"People are willing to give up their individual needs, to serve the group's needs, *if* they trust us," Paul Wicks says of PatientsLikeMe. "So

---

* The model being referenced here, created by Charles H. Green, names how the elements come together to formulate a "trust equation." He said three of these elements— credibility (can you), reliability (will you), and intimacy (we get where each other is coming from)—are in the numerator, yet the one that can undercut it all is in the denominator, which is self-orientation (will you choose yourself over us?). See Trusted Advisor, "Understanding the Trust Equation," http://trustedadvisor.com/why-trust-matters/understanding -trust/understanding-the-trust-equation.

the imperative lies with us, the organization [to earn that trust], and so we think a lot [about] the reasons people participate."

"You know how many people would be fooled if we were to say one thing and do another?" asks Jamie. "About five percent! Because you can't fool people at scale." So, he argues, "why not just share what it [the economic model] is? Be thoughtful about it, of course, but then let people make their decision about whether it's worth it to them?"

It's a good question, isn't it? Why not tussle through the trade-offs, publicly share the reasoning, and trust each other in the process? Wouldn't that be treating all parties as equals?

"This is really about a social contract," says Jamie, "where everyone understands the exchange they're making, and the benefit they're getting, and what are the economics at play."

"My job is to make sure that we fulfill the promise that we've set up," says Paul. PatientsLikeMe lost a $4 million deal, for example, because it wouldn't violate a core principle that guaranteed that patients would be able to see the results of all research. "If we do what we say, and we do it every day, it adds up to us being a firm that does what we say we will do. We earn the right to do more. But if I sell an e-mail address to a company to spam herbal supplements or something like that, it would be a complete betrayal of what we promise." And, he adds, "We would implode, and we *should*."

When a community shares its failures, it shows how and what it is learning and thus demonstrates increased competence. When it shares its decision-making process, it reveals its values. When it makes a tough choice and shares its reasoning, it demonstrates its integrity. When a leader has demonstrated selflessness over a period of time, it shows he will reliably choose the community interest over his own.

STEPHEN LIVES ON

It used to be that after a sick patient died, so did her ability to contribute to a cure for her particular disease. What PatientsLikeMe aims to do is to enable the deceased to live on in the form of data, data that is shared and freely available. As Paul explains, "People will pass away, and one of the first things that their spouse or parent or other caregiver will do is go on PatientsLikeMe and update their profile to let us know that they've passed away. It's incredible in the midst of all that chaos that they're thinking, 'No, I want this data to count. I want this data to go into some sort of official record.'"

What do we all want? We want our lives—even in death—to have an impact.

Illness robs its sufferers of their power in the world. "It takes away your ability to be a husband, or a wife, or a parent," Jamie observes. "Steven and I talked about [this] a lot . . . He got to the end of his disease, increasingly locked in, and unable to communicate . . . and we reached a shared understanding of what it means to live."

Sharing, however, helps restore it, as Jamie observes: "Life is worth living as long as you make the lives of those around you better. And I think that a lot of our modern malaise in our hypercompetitive, capitalist-driven world is that while it is dramatically more efficient, the fundamental reward you get for making someone else's life better is broken for most of us. The chance to see your work directly help another human being is reduced."

"My brother died, but he lives on in his data," Jamie muses. All of Stephen's experiences, medicines, treatments, and progress (or lack thereof) are a part of a permanent record, one of thousands of members of PatientsLikeMe, so that even though he is no longer here in body, that data has a heartbeat, and it just keeps giving.

PART III

Denting

Acting as One, Meaningfully

The first two sections of this book have discussed the power of claiming your spot in the world and uniting with others to gain the power of us. These, combined, let you find and own your wild ideas. But this is not enough to make a dent. As the parable goes: when five frogs are sitting on a log, and one decides to jump, *how many frogs are now sitting on the log?* Still five, of course, because deciding and acting are not the same thing. Part III of *The Power of Onlyness* shows how to get that crucial action of many while still honoring individual purpose.

First, to scale an idea, you'll need to know how to galvanize a crowd to care enough to act. Chapter 7 examines what "following" means as you direct people toward an outcome. It will recount the story of an Austrian who survived the Holocaust only to fight a long battle to get reparations, analyze how Occupy Wall Street wasn't concerned enough

with outcomes, and discuss how Black Lives Matter is effective in helping those who don't have the lived experience to still "get it."

Next, how do you get people to execute an idea as passionately as if
it were their own, acting on its behalf without being told what to do?
Chapter 8 explains how both Ushahidi and 100Kin10 enabled many to
act together to build new solutions, and it shows how much clarity of
shared purpose can keep chaos from happening through the case of
the Boston Bomber situation.

Chapter 9 considers the design principles that can unlock the onlyness of many, through the story of Foldit, identifying the three key
mechanisms that release the power of wild ideas in any system, even
more traditional organizations.

These three chapters show how to make ideas mighty enough to
actually make a dent.

CHAPTER 7

Galvanizing Many to Care (Enough to Act)

There's something happening here. What it is ain't
exactly clear . . . There's battle lines being drawn. Nobody's
right, if everybody's wrong . . . A thousand people in the
street. Singing songs and carrying signs. Mostly say,
hooray for our side.

—BUFFALO SPRINGFIELD

ACCOUNTABILITY FOR SNCF

"As soon as the train started in the morning and we were out of Paris for about half an hour or so, we decided our task was to pry these bars apart," recalled Leo Bretholz.[1]

If Leo had remained on that train from the Paris suburb of Drancy, seventy-two hours later he would have been at Auschwitz,[2] the German concentration camp where gas chambers had already been operating for seven months. But he managed to escape the train, which was owned and operated by SNCF, the French national railroad, which transported people a thousand miles eastward to their deaths that day. SNCF was paid for this service by the kilometer and by the passenger.

From March 1942 to July 1944, more than 75,000 Jews were placed on SNCF trains. Only 2,000 survived; of those who perished, 2,000 were children under six years old, 6,000 under thirteen, and the youngest only a few days old.*

Decades later, Leo, who had since emigrated to Baltimore, was convinced by a writer friend to tell his story, resulting in *Leap into Darkness*, his riveting 1999 memoir documenting his seven escapes from the Nazis as well as SNCF's role in the war. Writing the book was the beginning of Leo's journey to look more closely not only at his own history and experiences but his visions and hopes—a process that ultimately led him to seek justice by asking SNCF to take responsibility for its actions nearly seventy years earlier. The court battle that ensued lasted more than ten years, but even then, justice was still not served.

Then Leo harnessed the help of 160,000 "friends" who signed an online petition, and together they acted as a lever of change. They pressed SNCF to admit its wrongdoing and pay reparations of $60 million. What exactly inspired this outcome? To make a big dent, you must galvanize action: You must get others so excited or concerned about an idea that they, too, become motivated to want to do something about it by sharing its purpose.

GETTING OTHERS TO CARE

Leo's book laid bare the key facts: that the Red Cross had unsuccessfully attempted to intervene on behalf of the passengers, asking for better conditions from SNCF—for water to be provided, among other things. Many prisoners died just from the journey. More disturbing still, it was SNCF, and not the Germans, that chose cattle cars over

---

* There are two different fact sets for how many children were killed as a result of SNCF deciding to put children on the trains. In Leo's book, the numbers are near eight thousand. In official US Senate testimony, Leo mentions eleven thousand children, a number in agreement with the 2010 based-on-true-life French film *La Rafle* [*The Round Up*], directed by Roselyne Bosch. If "children" refers to anyone under eighteen, a larger number would make sense.

passenger cars, established the conditions in those cars, and decided to include young children in the transport.

In 2000—soon after *Leap into Darkness*'s publication, though not directly related to it—an unsuccessful class action lawsuit was filed to compel the SNCF to take responsibility for its actions and to pay reparations. In 2008, Raphael "Rafi" Prober, a young attorney at Akin Gump Strauss Hauer & Feld, began working on Leo's case.* Rafi's boss, who had recruited him, had been handling the matter pro bono since 2006.

Rafi "quickly realized how interested I was in the subject."

Some of that interest was personal. "My own grandfather was from Lithuania—he left for the United States before the outbreak of the war . . . his entire family was exterminated with the exception of his youngest sister, my great aunt." But Rafi was also deeply moved by Leo's own story. "[T]he pleadings and court filings and all that stuff . . . was incredibly compelling," he recalls, "but it wasn't until I read Leo's story, in his own words, that it overtook me . . . How could this have happened? How could no one know it? It boggled my mind.

"I [also saw] my grandfather in Leo, who came to the US by himself at age thirteen, went to night school, delivered newspapers, made his way . . . It was just crazy what [Leo] went through, yet he was engaged in the world, full of joy, a great guy, really sweet. Just like my grandfather."

Though most Nazi-era collaborators had taken steps to make reparations, SNCF argued, in case after case, that the law didn't apply to them. Although the panel of the United States Court of Appeals for the Second Circuit could not compel amends, it did write a strong opinion, stating, "[Despite] the evil actions of the French national railroad's former private masters in knowingly transporting thousands to death

---

* The lawsuit that Rafi first joined had been begun in 2000 by Stephen Rodd and Harriet Tamen on behalf of Holocaust victims of the SNCF. See *Abrams v. Société Nationale des Chemins de Fer Français*, docket no. 01-9442 (2d Cir. 2003), http://caselaw.findlaw.com /us-2nd-circuit/1046866.html. That lawsuit never got to resolution.

camps during WWII, [they] are not susceptible to legal redress in federal court today."[3] SNCF avoided legal liability by claiming in France that it was a commercial entity and then maintaining in the United States that it was French-government owned. Despite various dead ends, the legal team and the victims persisted.

Then, in April 2009, President Obama announced a high-speed rail initiative, earmarking $8 billion ($53 billion more was allotted in 2011)[4] for an update to US railroad infrastructure. Soon after, Rafi received a Google alert when SNCF responded to the request for proposal, declaring its interest in being considered for the project.

"I remember thinking, *that's crazy!*" Rafi recalls. "How could they come . . . looking for [US] tax dollars from survivors like Leo whom they have refused to acknowledge, let alone provide reparations?"

Rafi and his colleagues began to work, state by state, for laws requiring SNCF to release its war-era records into the public domain, which they believed might help them prove that the company's motivation was wartime profit and not capitulation to the Germans.

In 2010, their efforts resulted in some small dents: The SNCF disclosed one million pages of documents in order to bid on a rail contract, commissioned an independent report on SNCF actions called the Bachelier report,* and, for the first time since the war, it finally "apologized."† In doing so, however, it accepted no responsibility for its actions, arguing, "We were under inescapable duress, so anger [should be directed] at the

---

* The Bachelier report supports Leo's argument and refutes SNCF's assertions that it was an innocent victim. The report specifically says the SNCF negotiated for months the terms and fees they would earn. "The company would be remunerated at *the* best price for its services," for its 1,000 locomotives and 35,000 cars. If you're being *forced* to do something, do you normally have the capacity to negotiate terms? See Rémi Rouquette, "The French Administrative Court's Rulings on Compensation Claims Brought by Jewish Survivors of World War II," *Maryland Journal of International Law* 25, no. 1 (2010): http://digitalcommons.law.umaryland.edu/mjil/vol25/iss1/14.

† The *Los Angeles Times* took umbrage, noting that "the apology was apparently not prompted by regret." Instead, they wrote in their opinion piece, "It seems to have been spurred by the company's desire to win multibillion-dollar high-speed rail contracts in California and Florida." "Echoes of the Holocaust," editorial, November 20, 2010, http://articles.latimes.com/2010/nov/20/opinion/la-ed-apology-20101120.

Nazis, not SNCF," an SNCF executive remarked, continuing, "Making a statement of regret doesn't mean we're guilty. Making a statement of regret means we acknowledge the pain that lives on."[5]

In 2012, Leo and hundreds of others who were similarly affected launched an increasingly aggressive campaign to pass the federal Holocaust Rail Justice Act,* to allow Holocaust victims and their families to sue SNCF in American courts. He argued persuasively at a US Senate hearing: "As it was during the Holocaust for SNCF, so it is now—all about money."

Still, despite a compelling argument, hard work by very talented people, and hundreds of thousands of dollars' worth of pro bono legal effort, swaying SNCF was proving to be difficult. What convinced a Fortune Global 500 company—with its army of PR people and lawyers that had successfully fought against paying reparations for nearly seventy years—to change its position?

Was it a matter of Leo and his friends having a just cause? No, because they already *had* the just cause. Was it the effect of a moving story, told well, that caused people to want to support Leo? Since they already had a book and other such artifacts, it wasn't just the story that mattered. Or was this, as some might argue, a case of "clicktivism," in which disconnected Internet surfers simply get to feel virtuous by clicking a mouse? Probably not; we can all point to cases where someone clicked but nothing happened. Yet in Leo's case, something did.

Instead, it reminds us of the power of onlyness—how an extended purposeful community formed around an idea has a lever to move the world. Which is exactly what happened next—starting with Rafi Prober's friend Matthew Slutsky.

THE PETITION

Matthew was a very early employee of Change.org, having joined the organization in 2009. His route to Change was through politics and

---

* Note that this bill had been introduced as early as 2006, but was subsequently renamed the Holocaust Rail Justice Act.

policy advocacy. On the day he finished college in 2004, he drove to Iowa to work for presidential candidate John Kerry. Eight weeks and 14,000 miles later, he realized that "this is not the thing that has the power to change the world." He spent the next four years doing non-profit advocacy, then joined others working to make opposition to Darfur's genocide one of the most visible social movements of the decade, ensuring "celebrities, politicians, and every possible dollar was available to solve the problem." Despite engaging influential people like George Clooney, Oprah Winfrey, and the secretary of state, "the situation didn't really change," Matthew said.

By the time he joined Change's CEO, Ben Rattray, for dinner in June 2009, Matthew was wondering, "What *exactly* would it take to create effective social change?" He had directly experienced various forms of traditional power—money, influential people, title, rank, authority, expertise—and found them wanting.

Despite the promise of the Internet to democratize access to power, it often didn't allow entry to new ideas or to people who appeared "wild" to the rest of the world. On his way home after his meal with Rattray, Matthew called his mother and told her, "If just 10 percent of what this man just told me comes to be, I think this is going to be remarkable." Ben's vision for Change was "just like Amazon is for books, YouTube is for videos. There needs to be one place for change makers."

Rafi recalls learning of Change in its early days as "kind of a 'meh' moment." But then, one day in 2013, after years on Leo's case, he found himself wondering what stone had been left unturned. He asked a few close friends, including Matthew, "What else [do you think] can be done?" Matthew replied, "I think it could have a lot of resonance [on Change.org]."

So Rafi visited Leo, now ninety-two, explained what an online petition was, and showed him the Change.org site. "Leo was totally on the ball with technology," Rafi recalls. "He would send articles around and include attachments, so he used the Internet pretty actively. But he had never heard of Change [and was] skeptical as to what it could do.

He wondered how some strangers could help, when legislators and lawmakers hadn't been able to."

Nevertheless, Rafi convinced him to try it and worked to capture Leo's voice and passion—"He was a force of life"—in the petition. Leo approved it, and on January 31, 2014, it launched on Change's website.[6]

"Enough is enough," it began.

> It is time for SNCF to be held accountable for its active role in the Holocaust. Tell SNCF and its American subsidiary, Keolis, that as they seek to expand their business in the United States—with many projects funded by the tax dollars of the very survivors who were deported toward the death camps on SNCF trains—they must pay reparations to these Holocaust survivors and their families. It is simply unconscionable that SNCF's American subsidiary is now competing to build and operate the light-rail Purple Line in my home state of Maryland—valued at more than $6 billion and one of the single biggest contracts in state history—while refusing to be held accountable . . . I am almost 93 years old and we don't have much time left to act. Please, help me and my fellow survivors see justice from SNCF within our lifetimes.

CHANGE

By posting an online petition, a change agent can engage a community of people who have a propensity to care about a given topic and who see their clicks as a way to join in and make a dent. By the time it had sponsored Leo's cause, Change had started to become a site of self-selected, self-motivated activists that had come to the site for a particular issue but then remained to see if they could lend their efforts toward other campaigns as a lever to move the world.[7] Together, its members were able to spread and grow ideas from person to person, virally.

Virality is the power of an object—a video, petition, image, or meme—to be circulated rapidly and widely on the Internet. If a common factor in viral topics can be identified, it is a core idea that resonates. Memes based on humor or love tend to be resonant, as is any subject that is specifically meaningful to a large group of people. As Tara Hunt, a pioneer of online and social marketing, advises, virality occurs naturally when people *want* to share something that resonates and matters to them. Despite the efforts of many people (and organizations wanting to establish brands), trying to figure out how to make something go viral can't just be forced to happen. Companies, for example, want to tie specific online promotions to a product or a market expansion. But product launches are often clearly more about money than real meaning. What *does* work is an effort that isn't obviously based in a transaction. Instead of thinking, *We'll know we're successful if x number of people click*, true success is about finding and uniting individuals with similar concerns.* A good commercial example would be what the brand Dove has done to connect with consumers around the resonant issue of how limiting the current definition of beauty is.†

---

* Many of us don't make a distinction between awareness and outcomes. But since we care about the dent, success needs to be measured by demonstrable proof that you have achieved goals. Researchers Erin Gore and Emma Bloomfield draw the distinction between "signaling metrics" that measure progress, such as Facebook "likes," and "confirming metrics," such as statements by influential decision makers. See "Signaling and Confirming," *Stanford Social Innovation Review*, November 25, 2014, https://ssir.org/articles/entry/signaling_and_confirming.

† Dove started a conversation in 2004 based on the insight that only 2 percent of women see themselves as beautiful. Year after year, they built on that idea. In 2013, there was "Dove Real Beauty Sketches" where women were asked to describe themselves to a forensic sketch artist (who couldn't see his subjects), and then the same women were described by strangers whom they had met the previous day. When the two images are compared, the stranger's characterization was both more flattering and more accurate. See Tanzina Vega, "Ad About Women's Self-Image Creates a Sensation," *New York Times*, April 18, 2013, http://www.nytimes.com/2013/04/19/business/media/dove-ad-on-womens-self-image-creates-an-online-sensation.html.

This spreading and scaling effect of viral content depends on both strong and weak ties within networks.* Strong ties are useful to establish trust between people so they can lean on each other and build ideas, and organize plans together. Many of the stories in this book thus far focus on strong ties, where people find one another, agree to a shared construct, and, as a result, have better information and support. But if you were limited to strong ties alone, it would mean information would remain with those deeply in the know. Weak ties are formed when people who are on the periphery get exposed—for example, by seeing a post or picture on Facebook or Snapchat—to a new idea. Without strong ties, there is no *there* there, no community. But without weak ties, new and novel ideas don't spread.

Change.org's viral efforts on behalf of Leo didn't take long to work their magic. Whenever someone signs a petition on its site, a targeted policy maker or decision maker receives an e-mail notification. In the case of Leo's petition, the governor of Maryland was urged to support then-pending legislation in Maryland that would have precluded SNCF and its subsidiaries from competing for a contract before reparations had been paid.

Today, Change.org sponsors some twenty thousand new petitions every month. A quick glance at its website reveals that some petitions have been effective with as few as several dozen signatures, for narrowly local issues, to as many as one million or more for big national or international campaigns. Part of what makes Change's petitions so effective is that they function like a spotlight, focusing attention on one key person, on a single issue, with one specific demand. The speed, scale, and efficiency of the Internet in reaching members, accumulating signatures, and sending targeted e-mails is, of course, a key factor

---

* Mark Granovetter, a sociology professor at Stanford University, investigates the way people, social networks, and social institutions interact. His research shows the value of weak ties. See Mark S. Granovetter, "The Strength of Weak Ties," *American Journal of Sociology* 78, no. 6 (May 1973): 1360–80, https://sociology.stanford.edu/sites/default /files/publications/the_strength_of_weak_ties_and_exch_w-gans.pdf.

in making this process work so effectively. Previously, any such campaign would require months of door-to-door petitioning and money for envelopes and postage.

On February 4, 2014, four days after the petition on behalf of Leo's cause launched, Rafi called to tell him, "This Change thing . . . it's moving," and asked him to guess how many people had signed the petition. "People are busy," Leo replied. "I don't know, fifty?" "What if I told you it was fifty thousand?" Rafi answered.

"The Change thing was incredible to watch as it took off around the world," Rafi recalls. "It just affirmed for us . . . that these people . . . deserve justice. The petition represented a culmination, an affirmation that we're not crazy. That they deserve justice, and now is the time." In the end, 165,196 people added their names to the petition.

And, not long after, reparations were paid.

Sadly, on March 8, 2014, Leo passed away in his sleep, two nights before he was to testify before the Maryland State Legislature. Rafi was devastated by the news, but Leo's friend Rosette Goldstein, who was also a Holocaust survivor and SNCF victim and had been involved in the reparations effort for many years, took over the mantle of the campaign and wrote a beautiful note to signers of the petition to inform them of Leo's passing.[8] In December 2014, she was able to declare the petition a victory. "Today, I write you with very positive news on our decades-long fight to hold SNCF accountable for its Holocaust-era atrocities. On Monday, representatives from the US and the French governments signed a binding agreement that will provide a . . . compensation fund."[9] The agreement awarded $60 million to survivors and $4 million in donations toward Holocaust education.

Despite evidence that online petitions have helped causes like Leo's and the Eagle Scouts' in chapter 2, they do have their detractors. Critics often dismiss such efforts as "clicktivism" or "slacktivism," producing a warm feeling in participants but ultimately achieving little while undermining more effective forms of activism. It's true that there are many pointless Internet campaigns, as well as those that have genuine value for which people do take action without achieving success. (Think of such well-known yet failed efforts as Bring Back Our Girls, which sought to recover two hundred young girls who had been kidnapped in Nigeria.) Ultimately, your odds in any crusade will improve if careful organizing work is done, the shared purpose is clear, and a player in the decision-making chain who can be influenced is identified.

LESSONS AND TAKEAWAYS

Galvanizing is the way to find real people, anywhere, who share your commitment to an idea and engage them to act. It is not done through the marketing of a meme or by manipulating people to do your bidding but by meaningful actions. The story of Leo, Rosette, and Rafi— and all the other community members in their cause—leaves us with three valuable lessons for success in mobilizing many to act as one:

1. Be clear and specific in the outcome you desire. It's not enough to be outraged that the Holocaust took place or angry that SNCF committed heinous acts for money. The goal is to engage people to act toward achieving a specific result. A related practical lesson is not to focus on how many tweets or signatures you can gather but on measures that will most effectively serve to tilt the future in your direction.
2. The champion's role—exemplified by Leo, Rafi, and Rosette—is to be a trusted disseminator of accurate information that lays out a passionate onlyness-based case for why a particular issue matters. Charisma or influence is less important than authenticity and relaying one's concrete experience.
3. Petitions and other galvanizing mechanisms are effective, but there is no substitute for the concerted organizing efforts of a core group. Would Leo's petition

have succeeded if he and others had not already docu-
mented the issue and determined who needed to be influenced and what success would look like? The deeper
the foundational work to support a cause, the better the
chance of enlisting supporters. The organizing ground-
work and clarified idea will precede a successful galvanizing campaign. It's as though all the work of organizing
"loosens the cap" that a galvanizing effort pops free. A
petition (or phone bank or rally or protest) is not the first
line of action; it's the culminating tool or step.

These guidelines can be applied in other domains as well. If, for
example, you're an entrepreneur attempting to get others to understand your new market idea, you would begin by focusing on a specific
outcome, such as a near-term understanding by potential consumers
of what you want to achieve. The second lesson would also be valid:
You want to be a credible source of information that tells the story of
why your project matters. (If, for example, you're selling sugar-free
water, you wouldn't focus on the product itself but would stress the
core health issue that many care about and more should care about of
getting sugars out of foods.) Your personal integrity and its meaning
are what sell an idea, not some meme or its popularity. Third, you'd
take your time in drawing in many people until you had formed strong
ties and done the foundational work of establishing who should care,
who does care, what the shared action is, what objections need to be
taken into account, and so on. And then you'd make your idea spread-
able in venues where it can reach people via weak ties.

Technology-enabled advocacy sites are often credited as the key to
bringing about change. Of course, they deserve a great deal of credit,
given the leverage they now enable. But technology itself is too often
viewed as if it's the actual source of scale—the transformative agent
changing you and yours into the empowered collective you—when
really it's two pieces that, combined, make a whole. Technology is the

enabling agent; what actually creates the binding scale are the purpose, meaning, and shared values that are inherent in onlyness. Just as technology is displacing jobs, it is also opening up ways to connect and create value.

While this story showcased Change.org, it is more transferable than any one petitions site. Because things will change. It could be that people will learn to tune out digital campaigns like this, or that the business model of Change.org is unsustainable. But the point is that it's not about the tools themselves, but how people—you—come to use any tool like the one described in an intentional series of practices to galvanize many to care.

Purposefully acting toward solutions focuses attention productively. Imagine if, instead of telling the SNCF story and seeking justice and reparations on behalf of so many, Leo had held on to his grievances, convinced that he was unable to bring about change? Or worse yet, if he had stirred up community anger at SNCF but failed to direct that energy toward a practical solution—would reparations have been paid? The determining factors in the SNCF story remind us of the values of onlyness: that it gives you the agency to act, that it compels you to focus on what you can accomplish, and that it gives you the foundation on which to build a relational network of networks to move an idea from concept to reality.

BEING OCCUPIED WITH OUTCOMES?

You might have heard how they organized: Members signed up on a clipboard to share ideas at the twice-a-week forum called a General Assembly, held in Zuccotti Park, a privately owned public space at the center of New York City's Financial District. The crowd twinkled their fingers upward to show support for ideas, or wriggled them pointing downward to indicate it was time to move on. Consensus was reached by the multitude of dancing fingers.

This was how Occupy Wall Street (OWS) members collaborated in the fall of 2011. For all the global attention they received, though, did they actually galvanize a dent? Some argue that OWS has accomplished a great deal, especially as it raised awareness of inequality and the undue influence of corporations. Others believe it primarily fueled resentment and did not take any meaningful action.

One of the movement's central issues was that CEO pay was three hundred times the average pay of workers,[10] representing a system rigged to drive wealth to the top few rather than prosperity for all. While pay imbalance was not a new development (CEO compensation was four hundred times workers' average pay in 2000),[11] it became a flashpoint in the aftermath of the 2008 bank bailouts. OWS argued that the failure of the government to prosecute the banks showed that the democratic system had failed. There was also concern about the shortage of lucrative jobs to help pay off enormous student loans and the decline of unions.

Like the Arab Spring, and the Indignados movement in Spain, both of which also began in 2011, OWS involved people showing up in a public space to insist on change and not leaving until their demands were met. But, unlike the Arab Spring, which had a single, bold demand—the ousting of dictators—OWS had many.

HOW IT CAME TO BE

Though OWS launched on September 17, 2011, the planning had begun months earlier at *Adbusters*,* an anticonsumerism magazine based in Vancouver, Canada, that maintained that capitalism was failing society. The "Occupy" name originated in 2009 student-led protests against University of California budget cuts. *Adbusters* editor Micah White was present for these protests and joined in the occupation of UC Berkeley's Wheeler Hall.

It was in July of 2011 that White first proposed "occupying Wall Street" to *Adbusters*' 90,000-person US mailing list.† The promotion featured an image of a ballerina posing gracefully atop the Wall Street Charging Bull, captioned "What is our one demand?" and invited readers to determine what that demand should be, suggesting two specific options: a presidential commission charged with ending the influence of money in politics and a 1 percent "Robin Hood tax" on all financial transactions. It also proposed September 17, America's Constitution Day, as the date to launch the movement. After the July e-mail was sent to readers, some planning sessions were held by different groups in the New York City area, and although they were unable to agree on a demand, or a plan for convergence, OWS began on the September date as planned.

Three days later and three thousand miles away, White and Kalle Lasn, cofounder of *Adbusters*, were finishing a draft manifesto they intended to send to President Obama. They sought a tightening of banking regulations, a ban on high-frequency trading, the arrest of all the "financial fraudsters" responsible for the 2008 crash, and the formation of a presidential commission to investigate corruption in politics. "We will stay here in our encampment in Liberty Plaza"—another name for Zuccotti Park—"until you respond to our demands," the document

---

*   The group holds a special contempt for marketing. In their view, advertising exists to coerce and control. ("Adbusters," Activist Facts, accessed February 28, 2017, https://www.activistfacts.com/organizations/36-adbusters/.)

†   *Adbusters* has a circulation of 120,000. (Ibid.)

concluded. The manifesto was then e-mailed to Marisa Holmes, an "unofficial" OWS leader in Zuccotti Park.

"Micah, this is a wonderful draft," Holmes replied to the e-mailed manifesto on September 22.[12] "However, the General Assembly [in Zuccotti Park] is . . . drafting a statement [too]. It should be ready this afternoon." A week later, the "Declaration of the Occupation" was issued, presenting more a worldview than a list of demands. "We write so that all people who feel wronged by the corporate forces of the world can know that we are your allies . . . No true democracy is attainable when the process is determined by economic power." The six-hundred-word declaration went on to cite a number of grievances, blaming "corporate forces" for everything from poison in the food supply to cruelty to animals.

Instead of offering a specific proposal targeting a particular person with the authority to take action, they submitted a list of complaints, naming the vague target of capitalism as the culprit. To many in the park, the broadness of its target and the communal process by which it was arrived at was a virtue. Whether it was good or bad depends on one's perspective; if its goal was a specific action taken—a dent made—then OWS missed its opportunity. After all, there was already national consensus that pay inequity is bad for society; a Gallup survey found that 83 percent of Americans were dissatisfied with the economy. In June 2015, the *Atlantic* declared the rise in conversation about inequality as "the Triumph of Occupy Wall Street."[13] "The system is rigged" became a rallying cry for supporters of both the left and the right in the US presidential campaigns of Bernie Sanders and Donald Trump, respectively.*

But fully formulated ideas for how to address the problem were far more elusive.

In comparison, the Tea Party had a significant impact on the 2010 and 2012 elections, achieving specific policy legislation by putting elected

---

* Zeynep Tufekci, who studies technology's social impacts at the University of North Carolina, argues that the rise of Bernie Sanders was a direct result of OWS, and that OWS effects will continue to be seen in a Democratic insurgency in 2020. See "The Split," *New Republic*, June 14, 2016, https://newrepublic.com/article/133776/split.

officials in office within eighteen months of its 2009 inception. Many credit them with the Republican control of the House.[14] They galvanized many to act in alignment with a specific and shared set of principles, and getting specific politicians elected to accomplish those objectives was a measurable act of progress.

Did OWS follow the lessons of successful organizing discussed in connection with Leo's story? Did they identify a targeted person with a specific set of remedies? Although *Adbusters* attempted to unify the group around specific actions, those actions did not resonate with the community. It could be that the initial "community," their subscriber list, had not actually been aligned; they had nothing in common apart from their shared embrace of *Adbusters'* philosophy. Even in that, their stance was more adversarial than action oriented. In hindsight, asking ninety thousand individuals who lack a shared purpose to agree on "one demand" was not a recipe for convergence, rapid or otherwise. For these reasons, it's fair to say that while OWS stirred emotions, it failed to offer a clear enough path to make something new into reality.

Dissatisfaction with the status quo gives you a reason to leave, but not a place to go.

It's direction of purpose that lets you steer toward a horizon. To create a lasting change, you must have an idea of where, specifically, you want to see that change made, so you can start taking the appropriate steps to achieve it. What was OWS's ultimate destination on its path? Was it a candidate they supported? Lawrence Lessig, who teaches both law and ethics at Harvard Law School, was a candidate for president in 2016, though few were aware of it. He had helped establish the Creative Commons, central to online sharing resources (like Wikipedia) that allow crowds to build upon one another's work. In 2013, Lessig delivered an eloquent talk at TED,* arguing why money needs to be vanquished from the political process. He said we can't simply accept that changing our systems seems

---

* You can see the talk online: Lawrence Lessig, "We the People and the Republic We Must Reclaim," video, 18:19, TED.com, February 2013, https://www.ted.com/talks/lawrence_lessig_we_the_people_and_the_republic_we_must_reclaim.

hopeless. "It would be like you learning your son has terminal cancer. You wouldn't do nothing," Lessig said, "you would do everything."

Lessig's thinking seemed aligned with *Adbusters'* proposal to change the corrosive effect of money in politics. When he announced his candidacy in August of 2015, his campaign seemed tailor-made to appeal to *Adbusters* and the OWS movement. But faced with a possible advocate to address the problems they seemed eager to solve, they mostly responded with silence.

Whatever OWS's limitations, its activism may have indirectly benefited groups. For example, Strike Debt, Rolling Jubilee, and Debt Collective are tackling America's $1.3 trillion college-debt conundrum by buying back student debt for pennies on the dollar and forgiving it. The Corinthian Fifteen forced the closure of the for-profit Corinthian College amid claims of deceptive marketing and deliberately steering students into high-cost loans.[15] In Seattle, progress has been made in the fifteen-dollar-per-hour minimum wage movement thanks to one of the original organizers of OWS.

If galvanizing merely creates resentment, not only will progress not take place but conditions can become even worse. Action, not anger, is what's needed. Use emotions to engage and organize, to advocate and advance what matters. Enable your fellow dent makers to channel their energies toward reaching a specific solution and toward becoming a part of that solution.

Some may argue that anger is a necessary motivator for change, but it is a very risky strategy, as evidenced by the outcomes of the Arab Spring, or the Brexit vote, or the 2016 US election. People in those societies may have felt they had little to lose, but in the end, they put many things they value at risk. Any situation is not so dire that it cannot get a lot worse. Anger is useful in clarifying your values, but use it in a way that values solutions. As John F. Kennedy once said, "Those who make peaceful revolution impossible will make violent revolution inevitable."

CAN YOU GALVANIZE THOSE WHO ARE NUMB?

When Dr. John Wilson was named president of Morehouse College, he visited CEOs of companies based in Atlanta to get to know the
local business community. He found that the leaders of Home Depot,
Delta Airlines, Coca-Cola, UPS, and AT&T all shared with him the same
unbelievable statistic as a sign that they each understood there was
a problem: 80 percent of young African American men in the city
wouldn't finish high school. After presenting this at the inaugural Platform Summit in Boston, he was asked if more people knowing that fact
could help provide a solution. Wilson answered no.[16] "We've gotten
numb to this story," Wilson observes. "It's just not news anymore."

If numbness prevails, an issue has effectively become a fait accom-
pli and no dent is possible. Numbness isn't a matter of people's not
understanding or caring but one of their not believing that change is
possible. So how do you make the impossible possible? What or who
needs to be galvanized? And how can this be done?

The answers to these questions are critical because they shape how
people get big dents to happen, together. As we've seen, making a big
dent means galvanizing action, which involves getting people to act as
one. But what if, to bring about the crucial change you seek, you need
more than just a core purposeful group? How do you get those who
have no sense of connection to the issue (but need to) to care? Galvanizing, in this case, means making an issue so relatable that a broad set
of people can "get it."

This was the strategy adopted by Black Lives Matter to achieve police reform in many cities.

"I LOVE YOU. I LOVE US. OUR LIVES MATTER."

Wilson's comments were made July 13, 2013. That same night, George
Zimmerman was acquitted for shooting seventeen-year-old Trayvon

Martin, the son of a middle-class Florida family who had gone to a local grocery store for Skittles and Arizona Iced Tea and never made it home.

Alicia Garza, a special projects director in the Oakland, California, office of the National Domestic Workers Alliance, was at a bar as news of the verdict broke. She turned to Facebook to see what her friends were saying, and posted, "Black people. I love you. I love us. Our lives matter."[17] Those eleven words resonated, earning a hundred likes within only a few hours.

One of Alicia's friends, Patrisse Cullors, the reinvestment director at the Ella Baker Center for Human Rights, added her own note of frustration and reshaped Garza's words into the hashtag #blacklivesmatter.*[18] Two days later they announced a new project to "visibilize what it means to be black" in America. While neither woman positioned it as such, the statement "black lives matter" contained an implied "too"—as much as anyone's.

Later, their friend Opal Tometi, a writer and immigration rights organizer in Brooklyn, offered to build them a social media platform.[19] The three women started to put things into motion around this idea to make a dent. People in Los Angeles who saw the hashtag on Facebook began to use it for the first time a few days later on a banner in a street march that read "J4TMLA [Justice for Trayvon Martin LA]." In tiny letters beneath it, they scrawled "#blacklivesmatter," as if in an afterthought.[20]

Something was stirring. What it would become, as the Buffalo Springfield lyrics that open this chapter suggest, was not exactly clear.

WHY WOULD YOU CARE?

The night of the Zimmerman verdict, people online quickly began debating such questions as whether the decision was another case of

---

* #blacklivesmatter, Black Lives Matter, and the acronym BLM are not quite synonyms. I use the three terms differently. #blacklivesmatter is the Twitter hashtag, useful for Twitter-related references. "Black Lives Matter" is the name of a chapter-based activist organization. I use BLM for the overall movement.

systemic racism or whether the issue was more one of gun violence, or if the young man was a "thug."

Like many parents, I related to the tragedy through that lens, e-mailing back and forth on the plane ride from the Platform conference with my then nine-year-old brown son. As Wilson's word rang in my ears (. . . *we've gotten numb to this story* . . .), I digested the news. At first, a tragedy like Trayvon Martin's death and the acquittal of his killer seemed not to be definitively an issue of one thing or another. There were so many new pieces of information and various and often conflicting opinions, so it seemed impossible to formulate an opinion about race and the role of race in this death.

Given that, some people can simply decide, "It's not my problem," especially when there's no shared history or heritage. It's impossible to care for everything,* and, on complicated issues, it's hard to navigate the volume of information to understand something so new to you. It's one thing to read about something, another to be engaged with it meaningfully.

So, how does one get many to care?

When an issue is predominantly defined by a demographic, whether it is region, age, color, gender, or religion, that factor can easily become a boundary to who will care and therefore act on it. Sociologists call this an "in-group" for whom, by definition, the need for a dent requires no explanation; every member of that group simply "gets it."

Onlyness offers a different path to unity—not by demographic but by purpose; it is a way to connect with people who have the same hopes and dreams but may not share a common history and experiences. This shift keeps the focus on the *idea*—not on competing personal *identities*—so it can grow mightier.

And it requires a different way to engage people to care.

---

* Research shows that we have a limited capacity for empathy. Just as even the most determined athlete cannot overcome the limits of the human body, we cannot escape the limits of our moral capabilities. More here by Adam Waytz: "No, You Can't Feel Sorry for Everyone," *Nautilus*, April 14, 2016, http://nautil.us/issue/35/boundaries/no-you-cant-feel-sorry-for-everyone.

Grow a Mightier Idea

A "follower" is often understood as someone who accedes to another's desires and intentions. This is actually an example of *passive following*. A passive follower is typically dependent on a leader for guidance or is a conformist by nature. But the role of follower need not be passive.* Following can also be an active act. When you apply your own independent thinking and believe that your actions can make a difference, then following is always an active choice, a social undertak-

---

* Robert E. Kelly has a followership-styles model worth looking at. His four-quadrant grid maps the types of followers against a passive-active spectrum and a critical-thinking (independent or dependent) spectrum. My take on following is not directly based on his but inspired by his thinking. See Robert Kelly, "In Praise of Followers," *Harvard Business Review* (November 1988), https://hbr.org/1988/11/in-praise-of-followers. His work, as well as that of Barbara Kellerman, argues something important: followership as more than a mindless slate on which others write. That's like saying our bodies are unaffected by what we ingest.

ing that declares, "I care, too." When followers are linked by shared purpose, they actively engage with an idea. Active followership can change what you consider possible, what new facts you seek out, and even the conversations you're willing to have about an idea. Joining in is showing you care, actively.

ACROSS STATES, NEW STORIES FORM A PATTERN

The #blacklivesmatter hashtag was not very active during its first year.* But over time, more stories of black men needlessly dying made the news.

In July 2014, a year after the Zimmerman verdict, Eric Garner died in New York City after police placed him in an illegal choke hold. He told them "I can't breathe" eleven times; the release of the video of the event was followed by public rallies and charges of police brutality. When no indictments were filed against the officers involved, 4.4 million tweets tagged with #icantbreathe were posted in a four-day period, keeping the nation's attention focused on the need for police accountability. The mounting number of incidents pointed to the systemic nature of the problem.

Weeks later, on August 9, police officer Darren Wilson shot and killed eighteen-year-old Michael Brown in Ferguson, Missouri, after which the #blacklivesmatter hashtag exploded. Together #blacklives matter and #Ferguson accounted for eighteen million tweets during the month. On November 22, Cleveland police shot twelve-year-old Tamir Rice, who died the following day. Rice had been playing with a toy gun. Then, on November 24, when the grand jury yielded no indictments in the death of Michael Brown, 3.4 million tweets with the hashtag were posted in a single day.[21] This was apparently the turning point at which the #blacklivesmatter hashtag broke through to the

---

* A Center for Media and Social Impact (CMSI) analysis lists only forty-eight public tweets in June 2014. See Deen Freelon, Charlton D. McIlwain, and Meredith D. Clark, *Beyond the Hashtags: #Ferguson, #Blacklivesmatter, and the Online Struggle for Offline Justice* (Washington, DC: Center for Media and Social Impact, February 2016), http://cmsimpact.org /resource/beyond-hashtags-ferguson-blacklivesmatter-online-struggle-offline-justice.

consciousness of the out-group. As 2014 ended, the horrifying nature of these stories, coming in quick succession, meant that a growing number of people beyond the in-group began to discern a pattern from what could have been perceived as isolated incidents. Outrage online reached even higher levels in April of the following year after twenty-five-year-old African American Freddie Gray was arrested, allegedly for having an alleged illegal switchblade,* and his spine was broken while he was being transported in a police van.

This attention from ordinary Americans raised visibility of the cause to the notice of journalists, who began to analyze whether black males face different odds than white ones. The *Guardian* reported young black Americans are twenty-one times more likely to be shot by police than white Americans.[22] The *Washington Post* reported "that of all the deaths by the hands of the police, the rate of death for unarmed young black men was five times higher than similarly unarmed white men."[23]

Those facts, reported by independent press, helped add clarity to the issues.

MAKING SENSE OF IT ALL

By 2015, more Americans of all races felt what many black Americans already believed to be true: police brutality was a serious issue. Twitter became to Black Lives Matter what television had been to the 1960s civil rights movement.† In photos, videos, and Vines linked to the common hashtag, people witnessed these stories as they were taking place, exchanged ideas, changed their perspectives, and shared analysis, as a

---

* Both the possession of the knife as the cause of the arrest and the legality of the knife have been debated in court.

† The parallels are strong, from the number of people tuning in, to the newness of the medium. On March 7, 1965, forty-eight million Americans watched the scenes of protest in Selma, Alabama, on the news. Police dogs attacked activists, and bodies were bloodied for protesting peacefully. Martin Luther King Jr. said, "We will no longer let them use their clubs on us in the dark corners; we're going to make them do it in the glaring light of television."

critical mass of citizens began asking new questions. Forty point eight million tweets and more than 100,000 Web links with the hashtag were shared on Twitter alone in 2014 and 2015.[24] People were forging meaning, together, weaving the stories and facts into a national quilt of sadness.

Some have argued that social media—like Twitter—have the power to *create* a movement, while others maintain that they only *reflect* a movement. Both are possible. For our purposes, let's draw a distinction that matters. Two factors matter most in determining if a social media group is meaningful, says Dr. Jen Schradie,[25] an expert in social media and its relationship to social movements: First, a group must have a shared ideology; second, it must have strong ties. Without this relational context of purpose and trust, social media can easily become a platform for one person to yell loudly while others just tune it out.

Twitter-based hashtags as well as other social objects can facilitate sense making. Randi Gloss, a young entrepreneur, created and sold three thousand "And Counting" T-shirts that feature the names of those killed.* Beyoncé's song and related video "Formation" was among the most politically direct work she's done in her career,[26] with implicit commentary on police brutality, as well as Hurricane Katrina and the government's dismal response to it. Ta-Nehisi Coates created social objects with his June 2014 *Atlantic* feature article, "The Case for Reparations," and later with his 2015 book, *Between the World and Me*, which discussed police brutality and the conversation a black father has to have with his son.

All of these were forming a new picture, galvanizing Americans to ask new and urgent questions: Why was such violence taking place in the twenty-first century? Why was it disproportionately affecting black people? And what could be changed?

---

* Randi Gloss went to the fiftieth anniversary march on Washington in 2013 holding a sign with the names of five young black men who had died with a caption, "More than just black faces in tragic spaces." The sign drove many conversations during the march, prompting her to create a line of T-shirts for her company, Glossrags, whose text ends with "& . . .," suggesting the story won't be over until we act to change it.

This is how sense making works: people are galvanized to care, create an organizing system in which to place disparate facts, and reframe the topic to those who, at first, don't see themselves in the story and think therefore it's not theirs to "get."

FINDING NEW WAYS FORWARD

Lawrence Grandpre is the son of two black law enforcement officers—his mother was a state trooper, his father a Baltimore Police Department undercover narcotics agent who started serving after returning from Vietnam War service. As "blue" as those police accused of treating blacks unjustly, Lawrence applies all of this background, serving as policy director at a Baltimore-based think tank called Leaders of a Beautiful Struggle (LBS). LBS is one of many organizations doing work on social injustice, some long before the appearance of the #blacklives matter hashtag, including Black Youth Project 100, Dream Defenders, Justice League, Open Society Justice Initiative, Equal Justice Initiative, and Organization for Black Struggle.

In May of 2016, 61 percent of Americans believed more changes are needed to achieve racial equality, up from 46 percent in March of 2014 (up from 13 percent in 2010).[27] The majority of the US population is now galvanized to regard issues of racial inequality, especially equal treatment under the law, as important. Lawrence has concrete ideas to turn this heightened concern into action.

"The reality is . . . these are two families I belong to that are fighting," Lawrence says compassionately of his close ties to both Black Lives Matter and to the police. He describes how he used to worry about his father coming home safely, and the decisions cops have to make in the moment. "Without consequences of excessive force, the use of force keeps you safe enough to get you home. If there are no real consequences to killing young black men, then cops will use whatever it takes to get home."

It's clear what is needed: a reduction in police brutality by creating new accountability to one another, and to find a footing that readjusts the balance of power between citizens and those meant to protect them. "[I]t's not a fair fight. [Because] one has all the apparatus of power, all the resources, and the other does not. So, the issue has to be finding a way to change the power dynamics. It can't be about power over each other. As long as blacks are seen only as suspects, confidential informants, victims, and criminals—rather than what they are, the community that the police serve—we will never have a chance."

Lawrence believes that "the way to get cops to change their behavior is if you create a system that holds them accountable." In Maryland, he explains, "The chief has the power to discipline a police officer for a violation of protocol, from stealing money at a drug bust to shooting somebody. But . . . the police officer has the power to appeal that decision. That appeal goes to an internal trial board . . . made up entirely of other cops. That's where the phrase 'police policing police' comes from, because they can appeal the chief's decisions, and cops will usually side with other cops."

Lawrence understands that progress isn't easy. In 1857, Frederick Douglass, a human-rights leader and the first African American citizen to hold a high US government rank, said, "If there is no struggle, there is no progress."[28] Lawrence's group, LBS, seeks to embody Douglass's words to make progress. Formed in 2007 as a campus debate club at Towson University, the team won national recognition for using debate to frame real-world issues. LBS morphed into a political organization in 2009, advocating for specific policy changes.

One of LBS's main efforts is to get citizens on the police review board. "A ton of professions have professional standards boards that include civilians from the community, from elevator inspectors to architects," Lawrence observes. He views cops as a public good, and argues that they therefore have a higher need for community oversight.

Lawrence has shifted the conversation from managing an adversarial relationship to finding an entirely new way to advance progress by balancing the power equation. This, too, is a transferable lesson.

Framing and galvanizing are closely linked: You cannot galvanize without a common frame, so a reframing stage is critical. Just as Franklin Leonard and Samar Minallah Khan created new frames that fundamentally changed the dialogue, Lawrence is reshaping the frame of the conversation so it is shared by all sides.

Change the way people *relate* to one another, and you change the outcomes.

It's *how* you relate to something that causes attitudes to shift—not more arguments, and certainly not better reasoning. Think of how most of us try to convince someone else to adopt a different point of view: by developing a strong logical argument and presenting facts in support of it. But if an issue is "the way it is" for someone, no reasoning or well-argued set of facts will convince him otherwise. As Jonathan Haidt, social psychologist and author of *The Righteous Mind*, observes, "Reasoning is far less powerful than intuition."* If you are engaging a contentious topic like abortion, gay marriage, or income inequality with someone who is on the other side of the political spectrum, you are unlikely to shift his perspective with arguments. Instead, you need to first change how he relates to it.

"CAN'T JUST KEEP ON YELLING"

In December 2014, BLM activists were invited to the White House, where they told President Obama that they felt their voices were not being heard.[29] The president reportedly replied: "You're sitting in the Oval Office, talking to the President of the United States."[30] This was the moment at which the BLM representatives needed to realize that galvanizing is not just about pushing. To galvanize is not just to shock people out of their comfortable beliefs or to inspire existing believers

---

* For more, read Jonathan Haidt's article "Reasons Matter (When Intuitions Don't Object)," *Opinionator* (*New York Times* blog), October 7, 2012, https://opinionator.blogs.nytimes com/2012/10/07/reasons-matter-when-intuitions-dont-object/.

to effectively say "hooray for our side" but to inspire the collective group to take new action.

After that meeting, the steps BLM took over the course of the following months are what differentiate its program from OWS's. What BLM did successfully was to identify *specific* policy proposals to reduce police violence:

- Swift, legal, transparent investigations of shootings
- Tracking by race of those killed by police
- Demilitarization of local police forces
- Greater community accountability for police officers

In March 2015, the US Department of Justice released the results of a civil rights investigation that found a systemic pattern of civil rights violations by the Ferguson Police Department.[31] In May 2015, the Justice Department opened up a similar investigation of the Baltimore Police Department.[32] In addition, Brittany Packnett, one of the activists involved in both White House meetings, was appointed to the President's Task Force on 21st Century Policing. BLM also clearly influenced policies and politics at a local level. In Baltimore specifically, a bill passed in mid-2016 puts civilians on the review board for officers who are accused of crimes, the measure that LBS had been promoting. The bill also makes sweeping changes to the way police officers are hired and trained, so they know and belong to the community rather than serve as external or foreign stewards of it.[33] A revised Maryland Law Enforcement Officers' Bill of Rights, supported by the newly installed police chief, Kevin Davis, was also passed.

In February 2016, President Obama held another meeting, which was viewed as historic in that it included both traditional civil rights leaders of the 1960s and BLM activists. Those who attended spoke of making real progress,[34] writing, "It is in our actions that we find out who we are."

As BLM continues to make its dent, the movement has grown from a limited in-group aware of a problem to a mass of many people now galvanized to act.

What have we learned from these stories? The hashtag in BLM's case (as did the petition in Leo's) led to a shift in not only *who* was having the conversation but *what* the conversation was about. It certainly grew participation to include more than those who were directly experiencing the problem. It got people curious about the topic, inspiring them to learn more of the backstory. It united many people beyond individual history toward a common goal of equal justice for all. By specifying what needed to change, the group was able to translate its demands into concrete policies.

By uniting people through *ideas* rather than demographic *identity*, these denters gathered many together to galvanize action and even involved those who previously didn't "get it" as theirs to get.

CHAPTER 8

Commission to Own It

*If you want to build a ship, don't drum up people to collect
wood and don't assign them tasks and work, but rather teach
them to long for the endless immensity of the sea.*
—ANTOINE DE SAINT-EXUPÉRY

USHAHIDI: SPRING IT ON 'EM

"Without telling anyone [else] at the company, the three of us . . . just got to work, spec-ing some designs, thinking about how it would work, and building a [minimum viable] product," Brian Herbert, a senior-level engineer, says, describing the initiative of completely reinventing the core product of his then employer, Ushahidi. "We did [all] that," he explains, for many weeks and then "sprung it on the group."

Erik Hersman, a Ushahidi cofounder and then its director of operations, remembers getting an out-of-the-blue call, listening for an hour as the team described their dramatic redesign, and thinking, *You can't call that the same name [as the existing product]; it's so different.* Though Erik's initial reaction to the proposal was resoundingly negative, he tried to hide his disapproval, not wanting to squash his team's initiative and creativity or to signal that he didn't trust them or value their efforts, so he hedged: "Let me think about it."

Don't we all want the people around us to take responsibility for doing things that need to be done? In our personal lives, we expect our partners to share work, like making a fresh pot of coffee without having to be asked. As parents, we hope our kids will complete their homework without having to be tightly supervised. We want such behavior from our work colleagues, too, dreaming that they will take on new growth opportunities with the easy assurance of, "I've got this, don't worry."

However, much as we wish for these things, getting them to happen is hard. Getting those around us to take tasks on should not be a matter of delegating work, and dividing responsibilities into specific assignments. But, too often, that's *exactly* what it means, which is why most of us hate to manage or to be managed. The bigger the desired outcome, the more likely it is that the work involved will be divided into smaller and smaller tasks, to the point where, too often, they become devoid of any meaning, and people simply do just what is asked of them. In chapter 7, we saw how to galvanize many to care enough to act together. But in those cases, the goals were relatively straightforward actions: raising a hand, signing a name to a petition, taking a stand, having a conversation, sharing commitment with a friend. In those situations, the complexity lies not in the nature of the action but in how an idea becomes resonant to many. But how do you organize and get big outcomes when the necessary action is itself complex? For complex situations, is it possible to have many people co-own an outcome without it resulting in a chaotic and haphazard mess?

It *is*, but only if you know how to commission—entrust—people with purpose. As you do so, they not only *act* as if it's their ball to run with—it truly *is* their ball to run with. Let's see how they live this principle at Ushahidi, a Kenyan software company that collects crowdsourced information and then visualizes it on an interactive map.

CHECK-INS WITH A PURPOSE

After hanging up from the "radical redesign" call from Brian, Erik consulted with his Ushahidi cofounders, and by the following morning he had a new take on the situation.

"The problem we have as founders . . . is you have [an] idea of the way things could be, and . . . blinders on [for what else could be]," he explained. He realized that he ought to welcome disruption, even though it was now someone *else* doing the disrupting. So he told Brian, "Go for it. The worst thing that happens is we make a mistake. And if that happens, we'll fix it, together."

This turned out to be not just a random moment at Ushahidi but a core practice.

Ushahidi means "witness" in Swahili, and the company started in early January 2008, as violence erupted in response to Kenya's disputed presidential election. Erik, who had been raised in Sudan and Kenya by his missionary parents, saw reports of the strife while vacationing with his family in Athens, Georgia. As an active blogger on technology and society, he wondered, *How much of a hypocrite am I if I can't figure out how to use technology to help this situation?* He e-mailed his Nairobi technology colleagues, asking for ideas, but "all I heard was crickets." As he was trying to figure out who to partner with to do something, *anything*, he was inspired by a blog post he had read by Ory Okolloh, who on her blog, Kenyanpundit.com, proposed putting information about the election on a Google map: "It would be useful to have a record of this. For any future reconciliation process . . . the truth of what happened will have to first come out." Erik thought this concept could be applied to a real-time map that would enable people to navigate to safety, and started to consider who could turn that idea into reality.

He thought of other African technologists and bloggers whom he'd met at the previous year's TEDAfrica.* He called David Kobia, whom

---

* This event was the origin point of the TED Fellows program described in chapter 6.

he had just interviewed for a publication, and shared his concept. He didn't want to be rejected, so he added, "Let me send you some wire frames," though he didn't have anything sketched. After two hours of work, he clicked send at 9:00 p.m. and went to bed. The next morning, he found David had gone beyond responding to the conceptual idea to taking action; he had sent a link to a prototype. "I was like, 'holy crap,'" Erik recalls. "It was rudimentary; it was simple, but it was something that would work."

Meanwhile, Erik had also connected with Juliana Rotich, then a data analyst at Hewitt Associates in Chicago. At the time, she was in Eldoret, Kenya, on vacation, so she was in the middle of one of the flash points of violence. "The Kenya I grew up with is multiethnic and proud of it," she explains. "I went to a boarding school with kids from many different tribes, many countries, and so that was what drove my involvement in Ushahidi; the Kenya I know and that I was fighting for is one that celebrates many tribes, many languages, and the inherent strength of all of this [variety]."

Within forty-eight hours, Juliana, David, and Erik, along with Ory Okolloh and technology strategist Daudi Were, deployed David's rough but workable interactive mapping platform. Anyone could now text eyewitness accounts of their own situation (check-ins) and signal danger or urgent needs, and others could then avoid trouble or bring resources. Untold lives were saved.

Not so coincidentally, the goal of sharing information to protect people, which was the impetus for creating Ushahidi, was what led Brian Herbert to discover it.

OWN THE MISSION AS YOUR OWN

"I was in the Peace Corps in Kenya and had left the year of the election," Brian recalls, "but all my colleagues were still in the country . . . stranded . . . [because] no one expected [election] violence." When he could only find news feeds that contained secondhand information, he searched the Web, where he found Ushahidi. He told his friends in Kenya about it and, before long, started volunteering his skills to the project, writing code. Even after he left his full-time position in Atlanta in 2008 for an opportunity in Ghana, he continued his volunteer work in his spare time. A year later, Erik contacted him about the possibility of joining the organization full-time if some funding came through.

Erik explains that his hiring philosophy is based more on portfolio and passion than on traditional credentials on a résumé: "Show us what you've done and why you care, and we'll see what we can do together. The big thing about bringing people on was finding people who cared about the same thing; that means it's not about your ego but the bigger goals, the mission we're working on."

Does aligning purpose of person and organization matter? In the near term, as people bring more of themselves to the work, they more meaningfully engage their effort,* in comparison to people grinding out Task X for Wage Y.† In the medium horizon, when we lack engagement, it shows up in the bottom line. The cost of disengagement in business is $1 trillion annually in the United States alone, or about 6 percent of GDP.‡ And there's a profound longer-term effect. In a world of automation, where machines are replacing jobs, people will lose their sense of dignity, which comes from contributing that which only

---

* An average of 32 percent of the US workforce is engaged at work. That means that seven out of ten workers are not. "Employee Engagement in US Stagnant in 2015," Gallup.com, January 13, 2016, http://www.gallup.com/poll/188144/employee-engagement-stagnant-2015.aspx.

† In *Sensemaking in Organizations* (Thousand Oaks, CA: SAGE Publications, 1995), Karl Weick talks of how most work commoditizes the people involved. Citing Norbert Wiley, he writes: "Concrete human beings are no longer present. Selves are left behind and the structure implies a generic self, an interchangeable part as filler of space and follower of expectations but not the actual concrete individualized selves. This [work] relationship is [based on] abstract [people] rather than real [ones]." See also Norbert Wiley, "The Micro-Macro Problem in Social Theory," *Sociological Theory* 6, no. 2 (Autumn 1988): 254–61, https://www.jstor.org/stable/pdf/202119.pdf?seq=1#page_scan_tab_contents.

‡ Research done at the Martin Prosperity Institute provides three ways to assess the cost of disengagement: Nilofer Merchant and Darren Karn, *Onlyness: A Trillion Dollar Opportunity* (Toronto, ON: Martin Prosperity Institute, 2016), http://martinprosperity.org/media/Onlyness_A-Trillion-Dollar-Opportunity.pdf.

they can. It boils down to this: Aligning purpose of person and organization *is* powerful.

While Ushahidi's onboarding process includes a traditional "chaperone" to help new hires learn the systems, which digital tools to use, etc., the key to their onboarding is to get people to own the mission as their own. As Erik explains, "When anyone joins Ushahidi, we send them off to some conference to speak within the first few weeks. There's nothing like knowing you're going to be onstage talking about your new company to galvanize you into really learning the mission, and communicating that mission in your own words."

Inculcating mission is more important than sending the more experienced speakers to deliver the best on-brand message.

Brian came on board as a staff member in September of 2009. In early 2010, the Haiti earthquake and Washington, DC's "Snowmageddon" both struck. In each case, local members of Ushahidi's extended community took it upon themselves to launch Ushahidi-based solutions, enabling massive, crowdsourced local responses. In both cases, the sites became essential resources for many first responders, including government agencies. In the case of the Haiti earthquake, in which more than a hundred thousand people were feared dead, huge, centrally managed organizations like the Global Red Cross, FEMA, and the United States Marines used Ushahidi as a valuable hub to direct resources.[1] The Washington deployment was used to help clear streets.[2] Citizens and DC officials posted notifications of road blockages in real time, and available snowplows were dispatched to the worst-hit parts of the city so that resources were optimized based on those closest to the needs.

"WE WEREN'T IN LEADERSHIP, PER SE"

Ushahidi was built on the pothole theory, Erik shares: "The pothole theory basically states that you care about the pothole on *your* street, but not so much for the one two streets over . . . We [founders] were all from Kenya; this was our country. So when we saw that . . . things [were] falling apart, we collectively figured out what we could do to fix *our* local pot-hole." Now, with thirty-five full-time staff and a fluid ten-thousand-member volunteer community, their tool is deployed globally by people fixing *their* potholes. While you can't expect people to care about every-thing, with the right tools and rules, anyone can address their own needs.

Brian first got the idea of rewriting the core of the platform in late 2010. "We were . . . at ESRI [Environmental Systems Research Insti-tute] in Southern California, and brainstorming ideas on how to share information between maps. We couldn't find a way to do it with the way the [existing] software was built, so we just figured . . . let's find a whole new way . . . redesign it from the ground up to be more social. It was just an idea we—Brandon Rosage, and Evan Sims and I—had together. We weren't in leadership, per se."

Having developed a prototype, they then made that call to Erik. "The whole way Ushahidi worked was premised on self-starters who made stuff happen. We're always willing to take risks if it means better outcomes," said Brian. They were, effectively, fixing their own pothole.

The question that Erik struggled with after Brian's call was whether the pothole theory actually included the Ushahidi platform itself. Could he grant ownership of the company vision and the related prod-uct scope, and not just its execution, to Brian and his colleagues? And more to the point, could anyone at the company take this level of ini-tiative, or were certain things off-limits? Erik really wanted to believe that the pothole theory "applied to everything." And that's why, after sleeping on it, he green-lighted the work.

After eighteen months of development, Ushahidi released Brian's alternative crowdmap platform in parallel with the original. Rather than impose it on the user community, they let users choose what they preferred, and people chose the original. Although Brian's version is still active several years after its launch, most ongoing work is on the original Ushahidi platform.

"It didn't go as well as planned," Brian admits. "But we learned a lot from it. I really appreciated the freedom to experiment." Both he and

Erik describe it as a learning opportunity. Brian developed his skills as he realized his own vision. Heather Leson, a community specialist at Ushahidi, described how this "failure" helped make the whole team stronger: "Any time you are 'failing' in public, you're also proving that it's okay to take a risk and try new things." The wild message the entire organization receives is, *This company—including the mission, product, and strategy—is yours to shape.*

TEN HANDS ON THE STEERING WHEEL

Ushahidi's approach might seem, at first glance, a bit like ten hands on the steering wheel. "Go first, explain later," which is the way Ushahidi describes how it enables people to deploy without first seeking permission, sounds like a massive traffic accident just waiting to happen. How does this process actually work in practice?"

One key point is to keep people connected to one another and how the organization is delivering on its purpose.*

"We sit down every week with the whole team, saying, 'Here's what's going on,' so they know everything they can," Erik explains. Everyone in the core team must attend this meeting to share what's going on with them personally as well as in the organization. People share personal news about travels, their lives, or if they're sick, in addition to what they're working on. No one is allowed to sit quietly; participation matters. It is also a forum at which to ask questions, raise issues, listen to one another, and think together. "[I]t strengthens the organization if you can have people learn to problem-solve themselves, to come up with ideas themselves, and then figure out how to get there," Erik says. "That skill [to initiate and co-shape ideas] then proliferates as small groups and then bigger groups, and then the whole company."

---

* Tamara Erickson and Lynda Gratton researched what makes collaborative teams, and their number-one recommendation to leaders is to make sure team members know each other. See their article "8 Ways to Build Collaborative Teams," *Harvard Business Review*, November 2007, http://morris.lis.ntu.edu.tw/KM2016/wp-content/uploads/KM/W14 -1GrattonErickson2007.pdf.

The weekly meeting isn't the only way the Ushahidians stay connected. They take advantage of every possible channel of communication to do collaborative work. They used Skype channels in the early days of that platform, and wikis and blogs, then Google Chat and Google Hangouts, and more recently, Slack. Juliana explains that in its early days, with its founders geographically distant from one another, Ushahidi worked effectively despite remote engagement, which has deeply influenced how they collaborate now.

Still, proximity and in-person interactions do matter. Once a year, Ushahidi rents a place where the core team and partners assemble to chart their path for the coming year. Brian talks of the 2010 meeting where everyone converged on a big rented house in Miami. By planning together and thinking together they learn not only the ideas that are presented but why they matter to different members of the team. "[Even though] a month later, we're not on the [exact] path anymore," Brian comments, "we're still going toward the same light." It's clear that this proximity is the time to see the people behind the ideas. Erik's Flickr gallery of retreat pictures shows people in deep discussion around the dining room table, with people laughing, and it's easy to imagine Juliana telling some corny joke she loves.[3]

Ushahidi—both the organization and the platform—disrupts how most people think of organizing; its model enables the people who are closest to a given situation to report on it, share ideas with their peers, and to self-organize to solve anything. This self-directed approach assumes something fundamental: that people have the capacity to contribute based on what they distinctly see, that they can navigate their work in relationship to that of others without guidance from up top,* and that they want to help the organization flourish.

---

* Traditional organizations cascade strategy top-down, adding "substrategies" along the way. Ushahidi follows Henry Mintzberg's "emergent" strategy approach, which is built on a learning model where people can contribute their local information, compare notes, and shape what needs to happen, together. See Henry Mintzberg and James A. Waters, "Of Strategies, Deliberate and Emergent," *Strategic Management Journal* 6, no. 3 (July/September 1985): 257–72, http://onlinelibrary.wiley.com/doi/10.1002/smj.4250060306/full.

Ushahidi's approach taps into the capacity of onlyness. But what is the backbone structure of organization that makes that work?

STRUCTURELESS?

When people describe Ushahidi, they'll often use the term "structureless" as a way of describing how things are organized. But that term is a misnomer on multiple levels.

Likely, the word "structureless" is being used to differentiate between the more traditional centralized and decentralized organizational models.

In a centralized structure, the power and authority to plan and set direction remains at "the top." If Ushahidi were centralized, Erik and his cohorts would be directing others, as compared to what they do now: granting authority to many.

In a decentralized structure, authority, responsibility, and account-

Centralized Authority

ability are disseminated among various management levels, delegating others to own a piece of the whole. If Ushahidi were using the decentralized model, Erik et al. would find a champion in America and another in Haiti and have those people guide those pre-scoped operations.

Decentralized Authority

What Ushahidi is actually doing is more accurately called distributed work. Distributed work uses the network model, where each party, or "node," in network parlance, has the power to add its own value and to take its own share of responsibility. The distributed structure makes it possible for individual project groups like the Washington snowplow effort to do what they need to do without checking in with the Kenyan team, because the higher-level goal, or purpose, is inherent in every "node."

The key difference among these constructs is how information (and therefore power) flows.

The distributed construct is one where *any* person can share what

Distributed Authority

they know, based on their concrete experience, as an *equal* to the other players. With more than 90,000 deployments and with more than 6.5 million people participating, Ushahidi offers a compelling case study that a distributed model creates the kind of dents once reserved for governments and private-sector centralized organizations, and can actually create them faster.

There's a specific reason why calling Ushahidi "structureless" is a misnomer. It's that there is no such thing as a structureless group, as Jo Freeman has written in her essay "The Tyranny of Structurelessness." Any group of people that comes together for any purpose will inevitably structure themselves in one fashion or another to distribute power and organize. This is human nature. "Only if we refused to relate or interact on any basis whatsoever," Freeman wrote, "could we approximate structurelessness—and that is not the nature of a[ny] human group."*

---

* "The Tyranny of Structurelessness," JoFreeman.com, accessed February 28, 2017, http://www .jofreeman.com/joreen/tyranny.htm.

When a group doesn't have a well-defined set of rules, then people within that group will play their own power games to define what will be the unwritten set of rules to get things done. And that's the risk: Without clear and explicit guidelines for how things work, cliques and turf wars can easily dominate distributed work. But when the rules are well known, then each person knows how to navigate the terrain, trusts that, and therefore applies himself fully. Ushahidi's well-known rules, established through their weekly meetings, wiki-based communications, and open conversations, are what enable its extended team to understand both its mission and how best to work toward fulfilling it. With clear guidelines, the power to productively act toward shared goals is everyone's.

DYNAMIC COMMUNITIES NEED TRUST BRIDGES

"We've learned a hell of a lot on what it means to run a platform," Erik observed when assessing what made Ushahidi work. "[What] we realized was that technology was really only 10 percent of the solution. Everything else was the management of it, the communications of it." He came to recognize that "the common denominator for successful deployments that get the crowd involved . . . and using the platform for both sending and receiving information, is that they're run by or endorsed by people that others trust. These entities form what I call 'the trust bridge,' providing the necessary glue that brings credibility and trust so that others are willing to take part."[4]

As Erik views it, a trust bridge is a person who acts as a connector, advocating for rules and norms, and creating the magnetic bonds that hold otherwise disparate pieces together. A trust bridge enables you to manage people you don't see and can't control:* Because of the trust bridge, organizations or other people fill roles that you lack. A trust

---

*   Charles Handy, an organizational behavioral expert, was asked, "How do you manage people whom you do not see?" (or whom you cannot control or fire, because they may not be your employees). Handy's answer was simple: "By trusting them." Robert Cooper and Ayman Sawaf, *Executive EQ: Emotional Intelligence in Business*, (New York: Putnam, 1997).

bridge inspires people to bring everything they have—not just skills and resources but passion and energy and their purposeful commitment. Trust is central to onlyness-driven work. We've already discussed the trust equation in chapter 6. In Ushahidi, we revisit it to see it expressed fully. Trust functions both as an initial *enabler* and as a *moderator* over time.* Not only do the members of a group need to trust one another to begin a relationship but, once that relationship is established, they use a metric of trust to decide how much they will contribute. The deeper the level of trust, the more they will apply resources, put their own reputations on the line, and go all in. Trust is what allows you to commission people to take on more, because you have the cultural norms and related systems in place to let people help one another. Trust is central to how networked or distributed work works. And how onlyness scales.

The word "commission" has several very different meanings, but in its most basic definition, it is the act of entrusting a responsibility to someone else. The underlying assumption is that people fundamentally want to work toward solutions and that they have something to offer. It's what we need to do to bring out the best in anyone. When someone's capacity is wisely aligned with what they are commissioned to do, there's a good chance you can trust her to take responsibility. "People fundamentally want to help solve things, and now people can," Erik says, explaining his belief in each person's capacity.

LONE WOLVES

We've spent some time looking at the upside of the distributed, purpose-driven organization. But, sometimes, "the freedom to experiment" unfolds differently. Soon after Brian's crowdmap redesign project got going, a new engineer (let's call him "Lobo") came on board to build "Swift River," a data analytics tool for evaluating vast amounts of data very quickly.

---

* I wish I could take credit for this crisp insight. I was sharing this story at a meeting of the Martin Prosperity Institute with Adam Grant, bestselling author and Wharton professor, when he provided the enabler and moderator language.

"[Lobo] kept getting sidetracked by stuff," Erik recalls. Early signs hinted that he wasn't a great fit, but the Ushahidi approach to distributing responsibility meant giving him *more* rope. The situation came to a head a few weeks later, at the Ushahidi annual meeting, where it became clear that Lobo's goals weren't matched up to Ushahidi's purpose. Two days later, Erik learned that Lobo had formed a new company weeks before with someone he had recently hired at Ushahidi, suggesting that the pair had been working on it on Ushahidi time. This was over the line, and, as Erik puts it, "That's what made it easy to part ways."

Easy, but not simple. In the end, the team had to figure out how to deal with the situation, since the breakup occurred just as the firm made a major commitment around Swift River. This caused the expected angst, but reaffirmed what mattered at Ushahidi. Their policy was to give each person a lot of independence to deliver his or her best. It was clear that when Lobo defected from the Ushahidi team's shared purpose, the team could give up on him as well.

People like Lobo are "lone wolves"—individuals who choose to act alone. They are often quite talented, yet have a hard time choosing sides in the battle between their self-interest and the group's interest. Instead of reconciling these tensions so both needs can be met, they choose and act alone, on their self-interest, often selfishly. In many settings, this kind of behavior is often allowed, the thinking being that these people, despite being egocentric and temperamental, are needed. But this comes at a cost. The rest of the group will hold back, believing the situation is unfair,* and overall performance declines.

The risk lone wolves pose is that they are happy to exploit the rope they are given, using up valuable resources and time. They build what *they* value, not what is valuable to the shared purpose. By the

---

* The negative impact of what Wharton professor Adam Grant has labeled a "taker" is well researched. Robert Sutton wrote a book entirely focused on the topic, *The No Asshole Rule* (New York: Warner Business Books, 2007). Research done by Patrick Dunlop and Kibeom Lee confirms this; see "Workplace Deviance, Organizational Citizenship Behavior, and Business Unit Performance: The Bad Apples Do Spoil the Whole Barrel," *Journal of Organizational Behavior* 25, no. 1 (2004): 67–80.

way, their motivation is a clue: If someone is more worried about making themselves *rich* or *right* rather than serving the shared purpose, then it's not going to work out in a model that mobilizes people to act as one. Only when a group's shared identity wins over self-interest and selfishness can these lone wolves break through their egocentrism.*

Ushahidi's experience with "Lobo" offers two lessons. First, there will be times when you may be tempted to reconsider the effectiveness of the distributed, community model.† When you encounter a lone wolf trying to profit at your expense, the drawbacks of that model are highlighted, and the benefits are easy to forget. Remind yourself how the mission is best served when people feel commissioned to act, and how this fosters both creativity and speed.‡ People who abuse your collective trust will also serve as a negative example, reinforcing for the group the choices they must make and stick with.

The second, and perhaps more useful, lesson is this: Both the context of trust and shared purpose require a longer-than-normal time horizon, because they're predicated on relationships, which are a long game. Although we eventually discover whether an employee or group member is not going to work out, the unpredictability of relying on a specific person is hard to plan for and leaves the group vulnerable. It may be tempting, like parents who shelter kids from trying risky situations, to "play it safe" by micromanaging the lone wolf or assigning an important task to many people instead of counting on any one person. Recruiting

---

* Build a group ego by encouraging a single-minded focus on the goal. Bill Fischer and Andy Boynton advise this in their book, *Virtuoso Teams* (Harlow, UK: FT Press, 2009).

† Jeffrey Pfeffer has researched why in times of great change, people demand hierarchies. See Eilene Zimmerman, "The Case for Workplace Hierarchy," *Government Executive*, March 26, 2014, http://cdn.govexec.com/interstitial.html?v=2.1.1&rf=http%3A%2F%2F www.govexec.com%2Fexcellence%2Fpromising-practices%2F2014%2F03%2Fcase-work place-hierarchy%2F81301%2F. Jeff's work reminds me of William Golding's novel *The Lord of the Flies*, in which young boys are stranded on an island (the adults have died). Very soon, they call a meeting; they are trying to create order in a chaotic world.

‡ Talented people often get away with egocentric behavior because they are seen as offering an intelligence the group needs. Yet that notion discounts the evidence for a collective intelligence in the performance of human groups. See the research by Anita Williams Woolley et al.: "Evidence for a Collective Intelligence Factor in the Performance of Human Groups," *Science*, October 29, 2010, http://science.sciencemag.org/content/330/6004/686.

other members of the group when one person could do it would obviously waste valuable resources and slow things down. But worse, it would signal that people are not trusted, creating a predictable downward "holding back" spiral in which everyone trusts less and gives less.

Collaborative management—the management approach upon which Ushahidi draws—is often derided as an inefficient, "Kumbaya"-style technique, with everyone holding hands and singing ecstatically about a common cause. The reality is that network-based collaborative models simply operate differently. Most crucially, they share their mission/purpose widely to drive shared responsibility. There is great power in believing in people, and having them believe in something together. Research shows that only 5 percent of people in traditional organizations fully understand their fundamental strategy,[5] which means that all key decisions *must* be made at the top. Ushahidi makes certain that everyone, from its core team to its ten-thousand-member extended developer community, *knows* and *owns* the values that drive decisions.

It's simple, but it's not easy.

As of this writing, the Ushahidi platform has been deployed globally more than ninety thousand times in forty languages; about 80 percent of its usage is related to emergencies, elections, or crises. It was an invaluable tool when terrorists attacked Nairobi's Westgate Mall, after the Fukushima power plant disaster in Japan, and to track gender-based violence in Pakistan. But Ushahidi has grown flexible enough to accommodate almost any kind of geographical data. Every February, the Dutch launch a massive deployment to crowdmap which lakes and rivers are ice-skatable. There's also Brian's personal favorite use case, mapping North America's best hamburger joints.

WE ARE THE CHAMPIONS

Duncan Watts, author of *Everything Is Obvious *Once You Know the Answer*, once wrote that "building a successful life requires a deep conviction that you are the author of your own destiny. Building a successful society requires an equally deep conviction that no one's destiny is their own to write. Reconciling these seemingly contradictory ideas may be the most important social challenge of our time."[6] These words poetically capture what Ushahidi shows us in action.

Ushahidi resolves the tension by effectively saying, "*We* are the champions." Like Brian, we need not be "in leadership, per se" to champion an idea we think matters. And yet, when many people are commissioned to act as one, we can manage to fix many potholes—not because work was assigned but because "we" saw what was needed and "we" made it our mission to do it.

Juliana once took me on a Skype tour of her office, walking around with her laptop in hand. I saw the wing of an aircraft and cool, skylight-brightened areas. In the distance, I noticed the top of a trade show–style poster, maybe four feet wide and eight feet tall, the bottom hidden by desks. I asked what the sign said, reading out the part I could see, so she knew what I was referring to:

"You can do hard . . ." I read.

"Things," finishes Juliana.

YOU CAN DO HARD THINGS.

It's true. You—the plural form of you—can do hard things . . . together, when you engage the onlyness of many. Not by assigning tasks and work but instead by commissioning all the members of the community, in the words of Antoine de Saint-Exupéry, to "long for the endless immensity of the sea."

IDENTIFYING THE BOSTON BOMBER

Martin Richard was eight years old and in the second grade when he was killed. In one of the first pictures posted of him online, he has a big, toothy grin and is wearing a baseball cap. He died less than four feet from the second bomb explosion near the Boston Marathon's finish line on April 15, 2013. His mother reportedly leaned over him, begging him to live, as he bled to death. Later, the news would report that he was the middle of three children and had made a poster with a peace sign and the words "No more hurting people" for a school project.[7]

What drives people like Ushahidi's activists to step into complicated situations and try to help? The answer is obvious: When people are suffering, if you can ease their pain with something only you have to offer, then you will do so. If you have a group come together around that purpose, then all the better.

But, sometimes, "helping" can become hindering, as the next story shows. Until you understand how to commission action with mechanisms to ensure it is carried out effectively and responsibly, you might make a different dent than the one you're aiming for.

IF YOU HAVE PICTURES, UPLOAD THEM

The 117th running of the Boston Marathon was a day intended for joy and personal accomplishments, but became quite different after two brothers—terrorists—placed bombs near the finish line, killing 3 people and injuring 264.

Within a few hours of the explosions, thousands of people had started a conversation on Reddit to discuss how best to help the victims. Some participants identified locations of hospitals and clinics to donate blood, others listed housing for marathon runners from out of town who now needed shelter, and others mobilized for where to send

pizza to show support for the local officers who were working over-
time in the crisis.

One thread, started at 2:48 a.m. the following morning, asked, "If
you have any pictures from the attacks, upload them here" and included a button to click.[8] What followed was an Internet frenzy, with
a flurry of accusations and rampant speculation about the perpetrators
that falsely accused several men.

As photos were uploaded, Redditors began to circle people in the
crowd gathered that day who *seemed* suspicious. Wearing a backpack
made someone a target, as did being brown. One publication denounced
the effort as a racist "Where's Waldo?" exercise, as people were making assumptions based on stereotypes about what they thought terrorists looked like. But no one stopped it—not the originator, not the
volunteer moderator of the chain r/findbostonbombers, not Reddit,
not even the police.

Meanwhile, the media and everyone else were tuning in.

"It became almost its own beast," says Chris Ryves, one of the Reddit
moderators involved in the activities, in a documentary that was later
made about the social media manhunt, *The Thread.** He describes being
overwhelmed by the millions of users who descended on the subreddit
he managed, "Find Boston Bombers." These "helpers" proceeded to mis-
identify as suspects two innocent individuals whose images were subsequently featured on the front page of the *New York Post,* accompanied by
inaccurate reports that federal authorities were searching for them.[9]

Many observers believe these false accusations drove the FBI to
release photos of the official suspects (who would later be identified
as the Tsarnaev brothers). Though the Redditors had by this point

---

* Elaine Teng wrote a long-form essay about the documentary *The Thread*, which recon-
structed these five days between the bombing and the arrest of Dzhokhar Tsarnaev, the
surviving brother found guilty of all thirty charges in the case. Her piece ("A New Boston
Marathon Documentary Tries—and Fails—to Scare Us About the Internet," *New Republic,* April 9, 2015, https://newrepublic.com/article/121503/thread-new-documentary
-about-boston-marathon-bombing-review) is highly critical of the film because while it
gives some sense of the drama, it doesn't offer a clue as to the difference between what
helps and what hurts in galvanizing action.

proven their ability to be unhelpful, they continued to involve themselves in the case, choosing as their new assignment the identification of the men in the FBI photos. They seized on the resemblance between one of the men in the FBI photos and a twenty-two-year-old Brown University student, Sunil Tripathi. Someone on Twitter claimed he'd heard Tripathi's name on the Boston police scanner, and the story spread. Redditors noticed that Tripathi had gone silent on social media in recent weeks and concluded he was in hiding. Soon, Tripathi's parents were receiving threats from random sources, compounding the pain and fear of not knowing the whereabouts of their son. The people aiming to be "helpful" to the parents of Martin Richard were unwittingly harming another set of parents in the process. In fact, Tripathi was missing because he had committed suicide. (His body was eventually found in the Providence River.)

WHAT'S THE ISSUE AT STAKE HERE?

There are some who judge that the Reddit Boston Bomber situation was simply a matter of social media run amok, that well intentioned people were doing the right thing, and that if it hadn't been for the constant media attention to Reddit during the crisis, the matter wouldn't have been blown out of proportion. Some people view it is a lesson applicable only to journalism. Others argue that it is simply an example of the new "social" status quo, which is effectively a done deal.

Let's view the matter from the perspective of what we learned from Ushahidi's practice of commissioning many to act as one and its two main tenets. First, commissioning action requires developing and maintaining a shared vision of what success looks like. Second, it's important to cultivate people's motivations based on their onlyness, so they can and do build the capacity to own more of a given problem's scope, solutions to it, and how those solutions are executed. Did the Boston Bomber effort meet these criteria?

In some respects, it did. Redditors were motivated—they were trying to own the problem, solution, and execution. But did they have a clear idea of what constituted success in their campaign? What *would* success have looked like in this case? In one Reddit thread, it was defined as "if you have pictures, upload them," and in the other, it was "find the Boston Bombers." No one, however, raised the condition of success as *accurately* identifying the bomber by adding the cautionary note of, "If we get this wrong, we could cause harm to people emotionally and financially and even harm ourselves because we would violate the purpose by which we're collectively mobilized to act."

The specifics of the shared purpose and goal were not specific enough.

This account is a telling example of why people worry about distributed work. Their fear is that many things will be done, but not the *right* things. A key point you need to be clear about in galvanizing action is defining the constraints that will allow people to continue moving forward quickly, using their independent capacity, but still keep them heading toward the right horizon.

Without that crisp definition of what outcome equals success, task-oriented people—people like these Redditors—may do their best at many subtasks, but the collective outcome will be unsatisfactory. You might get lots of action, but not necessarily action that serves the common good. In the absence of agreed-upon goals, participants in a failed project might argue that it doesn't matter what the outcomes were, only that work had to be done. They might even judge naysayers as "bureaucracy," obstructing meaningful effort.

LEARNING WILL HAPPEN LATER

On April 19, 102 hours and 9 minutes after the first bomb went off, a tweet from the Boston police announced, "CAPTURED!!! The hunt is over. The search is done. The terror is over. And justice has won. Suspect in custody."[10] Which is a reminder that all of this went down very fast, so fast that the opportunity to reflect on what happened, and to tune the process, didn't come until after the experience was over.

On April 22, the general manager of Reddit, Erik Martin, took a moment to reflect and issue an apology:

> Though started with noble intentions, some of the activity on reddit fueled online witch hunts and dangerous speculation which spiraled into very negative consequences for innocent parties . . . One of the greatest strengths of decentralized, self-organizing groups is the ability to quickly incorporate feedback and adapt. Reddit was born in the Boston area. After this week, which showed the best and worst of reddit's potential, we hope that Boston will also be where reddit learns to be sensitive of its own power.[11]

With that statement, Martin and his Reddit team committed themselves to learning, an important factor in successful commissioning. You cannot guide people to new outcomes without being dedicated to educating yourselves as a team about what is working and what is not.

WHAT WE CAN LEARN FROM REDDIT'S FAILURES

What do we hope that Reddit learns that we can benefit from?

First is that good intentions alone do not necessarily generate good outcomes. The social media Boston Bomber manhunt is an example of mob behavior, much like the Gamergate incident. You don't want to say "do this" or "do that," which would be micromanaging. Rather, you must be clear and specific about what is an acceptable outcome.*

---

* Without clarity on desired outcomes, team members choose options based on unspoken yet differing assumptions. This sets the stage for a group failure, either because they end up unable to do something or because they turn the group leader into a "dictator" having to make the decision on the group's behalf. Often the response is to say the people failed, but really, the process failed the people. For more on this topic, see Bob Frisch's article, "When Teams Can't Decide," *Harvard Business Review*, November 2008, https://hbr.org /2008/11/when-teams-cant-decide.

Second, you want to have mechanisms in place in the event that the absolute worst happens. When Chris Ryves described the scene by saying, "It became almost its own beast," he was effectively acknowledging that he had no apparatus—either in terms of tools or social norms—for applying some much-needed brakes. Among the hundreds of comments on Reddit in the aftermath of the bombing, it was clear that some people were aware that the situation was getting out of hand and insisting it be stopped. Those in charge of a system have to design for the many ways the bus can veer off the road and how to steer it back on, and use known mechanisms to bring out the best in people, not the worst.

The more crowds act like herds, the less wise they are. Fear can direct groups to do damage. This is why the design of acceptable outcomes and norms matters so much.

100KIN10: TAKING ON THE IMPOSSIBLE

Talia Milgrom-Elcott put her two kids to bed and sat down to watch President Obama deliver his 2011 State of the Union address, in which he announced many initiatives and called on Congress to make his proposals their agenda for the upcoming legislative period. Talia listened intently, waiting for the president to reach the subject that most concerned her.

As the Carnegie Corporation program officer focused on improving teacher performance, she knew that teaching was a key factor in enabling real classroom learning.* She'd also seen an early draft of the report by the President's Council of Advisors on Science and Technology calling for training ten thousand new STEM—science, technology, engineering, and math—teachers in the following two years.

As Obama came around to the topic of education, Talia was shocked when he announced, "And over the next ten years . . . we want to prepare a hundred thousand new teachers in the field of science and technology, engineering, and math."

A hundred thousand?! Not ten thousand?! she wondered to herself. *How does one go about solving a challenge so big?*

Coincidentally, the day after Obama's January 2011 speech, Talia was scheduled to be at a meeting at Ideo's offices in New York City with twenty-eight organizations, including the Gates Foundation and Google, to discuss a policy proposal around teacher training. After the State of the Union address, however, the conference became about the 100K-in-ten-years gauntlet. Talia shared her own story with the group

---

* One study found that the top 10 percent of teachers impart three times as much learning as the worst 10 percent. Another suggests that if black students were taught by the top 25 percent of teachers, the achievement gap between blacks and whites would disappear. "How to Make a Good Teacher," *Economist*, June 11, 2016, http://www.economist.com /news/leaders/21700383-what-matters-schools-teachers-fortunately-teaching-can-be-taught -how-make-good.

and her conviction "that this divergent set of players could create a breakthrough solution."

As the meeting at Ideo progressed, the group felt uplifted by shared purpose. "Essentially, we decided to build what we called 'not your grandfather's coalition,'" Talia recalled, "because this network would be different. It would *demand* action. No 'sign on the dotted line' and then go back to business as usual."

This was the genesis of what would become 100Kin10, with each of the groups involved making commitments to one another and for a shared purpose. Teach for America signed up to train four thousand teachers; Talia's Carnegie team signed up to find other paths for applicants who couldn't get into the Teach for America program by raising a three-year fund of $20 million to help support their work. In the only program of its kind, the American Museum of Natural History signed up to train novice teachers, certifying them within the context of a museum by focusing on interactive surroundings instead of academics alone.

"Ensuring that people know how to solve complex and global challenges to excel in this time demands we understand more of math and science, and not only understand it but be in love with it," Talia explains. The field of education doesn't lack for ideas for how to fix the teacher problem, or advocates, or even funds committed to solving it, yet it remained persistently unsolved, year after year.

"I . . . effectively looked to the left, and then to the right and thought this [could be] yet another one of those calls . . . [which] no one was going to do anything about," Talia recalls. "Looking to the left and to the right was a story told of Moses in the Bible," she explains. "He sees a slave being beaten, and he looks to the left and to the right and sees nobody." Talia's mother, it turns out, was one of the first female rabbis in the United States. "Not that there were no people there, but nobody wanted to take action, to stand up for these others. And seeing that, Moses steps in. Whenever we see something that needs to be done, our job is not to wait, but to step in and step up."

Within a year, 100Kin10 had grown to a hundred committed partners, with each having a specific commitment tied to action. Some
provided funding, others put forward new ideas, and still others offered resources that had been previously untapped. At the end of the
first year, six thousand teachers were trained, the largest cumulative
effort that had ever been tracked. By the second year, the organization
had more committed partners, thanks to the early involved parties recruiting others they knew could commit to the same purpose. At the
end of that year, 12,412 teachers had been trained. This was beyond the
10,000 mark that Talia had previously considered "nearly impossible,"
but it was still off track to reach 100,000 in ten years. Rather than hope
for some miracle "later," Talia and her three colleagues—each, like
her, working on the side on this 100Kin10 initiative—started taking
stock of what would need to be done differently if the more ambitious
goal were to be met. "We thought maybe we should focus on making
the teaching of science *cooler*, maybe bring in help to make it more fun,"
Talia explains. "So we ran a national campaign. We partnered with
Cultivated Wit, an amazing team of former *Onion* writers and editors.
Some forty partners crowdfunded it. And CW designed an awesomely
funny campaign,[12] with animated video, and sharable bits, and quizzes
to create engagement. The site was well received, and cool."

Still, Talia felt that this strategy was not quite right, but she kept her
misgivings to herself, not wanting to dampen the enthusiasm and creativity. Soon, she couldn't stop herself from asking herself, and then
her team, a pointed question: "Are more undergraduates applying to
teachers' programs because of that campaign?" She was assured that
they were, but not in large enough numbers.

That's how she found herself, two years in, starting to doubt if they
would achieve their goal. Finding the solution to tough problems may
require taking a multitude of different paths. Sometimes it involves
researching what methods have proven successful and applying them.
For Talia's project, though, there was no "best practice" to emulate.
Sometimes it's a question of finding the expert who has the "right

answer." Again, the solution to Talia's problem was one that would have to go well beyond what any one individual had figured out thus far. Worse yet, she came to realize the real challenge was even greater than building a large teaching force in a relatively short period of time. There were many interrelated and complicated problems faced by teachers: "The lack of peer support that would let teachers learn what really works from other teachers, for example. Or, how unprepared teachers are when they enter the classroom; most of what is taught is the theory of teaching, not the practice of teaching."

To make a big dent on such a complex topic, Talia was coming to realize, wasn't a matter of organizing *more* effort by *more* people— everyone working harder—but, rather, of introducing a *new* effort through a *new* understanding of the issue involved.

The power wouldn't come from *doing* more but in *reimagining* what was possible.

As the effort reached its third year, Talia was awakening to the possibility that "we could reach the goal of providing a hundred thousand excellent STEM teachers but not solve the underlying issues that created the need originally. We would have accomplished the task without solving the problems."

She felt "a sinking realization that . . . we could be doing this wrong."

PUT A MAN ON THE MOON

Many people in Talia's situation would not have *expanded* the problem that concerned her but would have kept their focus on the key metric of a hundred thousand teachers. They wouldn't have questioned their strategy but put their efforts into doing *more* work and obtaining *more* resources.

That's a trap that is easy to fall into. Tough problems are tough for a reason.

Andrew Zolli, a well-known innovation thinker who has spent

years identifying and working with early-stage social entrepreneurs, has seen too many ambitious projects pitched with an identical heroic formula. "When presenting their ideas, aspiring world changers have all learned the same blueprint," he said. "They start by showing a morally outrageous circumstance—a decimated ecosystem, an impoverished community. Nobody with an ounce of empathy could resist being moved by the setup. Then, they proceed to show how, with some previously unknown, often modest technical intervention, *magically*, all the problems can be neatly solved. These solutions promise to 'solve' hunger with a smartphone, or 'solve' inequality with an app."

But complex problems aren't complex because of the lack of technology. Complex problems involve solving a series of interrelated sets of issues and changing the fundamental assumptions that have been baked into how things are. They require, in net, something quite different to "solve." Like the problem Talia faced, it will take something new. She began to notice the similarities between Obama's call for 100Kin10 and an earlier president's appeal to take bold action.

In 1961, John F. Kennedy gave a historic speech before a joint session of Congress that set the United States on a course to the moon, "not because [it was] easy, but because [it was] hard." Kennedy believed "we possess all the resources and talents necessary"[13] but acknowledged that the United States was years behind the Soviets in rocket technology. A trip to the moon was far beyond what the country was then capable of. It would require talents to create alloys that had never been created and go to speeds that had never been reached at temperatures that had never been experienced. No engineer or organization in the country knew how to do those things, yet a committed group of scientists and engineers came together to work on the project because in doing the hard things, they would learn the new things.*

---

* This is true even if some things we link to the space race, such as Tang, Teflon, and Velcro, were not so much developed for it as made famous by it. See "Are Tang, Teflon, and Velcro NASA Spinoffs?" NASA.gov, accessed February 28, 2017, http://www.nasa.gov /offices/ipp/home/myth_tang.html.

Talia and her team had up to this point been acting as harbor masters: helping inspire, gathering commitments, directing people toward the right horizons, and allocating resources. While those efforts were hugely important, they weren't going to commission the group to identify brand-new solutions that no individual party had yet to find.

Rather than organizing the work of 100Kin10 as a *collection* of people in a group doing work next to each other as they aim toward a clear goal, instead, they needed to morph into a *change-generating collaborative* in which the methods to reach the goal would all be designed and decided by the group.

It would mean commissioning themselves, trusting themselves to do the impossible.

We can see the similarities between 100Kin10 and Ushahidi: Both were seeking to achieve complex objectives through entrusting people to act as one. The Ushahidi community already existed as an organization, albeit a small one, which started with and matured around a shared purpose, aligned through a set of norms. For 100Kin10, there were significant differences in the member groups—some were funders, others were nonprofits looking for grants. Each had its own distinct mission and norms. Given those disparities, 100Kin10 would need more than a shared mission—it would need people to let go of what they do today and to formulate a new set of shared norms that would bring its goal within reach.

They decided to reinvent their entire engagement model.

FLIP THE RFP

Lynda Kennedy came to a 100Kin10 summit, hoping to learn what it would take for her organization's RFP [requests for proposals] to be accepted. Her organization, the Intrepid Sea, Air and Space Museum, had twice applied to join 100Kin10, but both times had been rejected. But something about its application—with its modest proposal to train just fifteen teachers—persuaded the review committee to invite vice president Lynda Kennedy and her team.[14]

"We see the same problem [over and over] in the foundations and grant-giving world," Saul Kaplan of the Business Innovation Factory observes.[15] In 2015, his own organization had been recruited by 100Kin10 to codesign a collaborative problem-solving approach. Their "work out loud" facilitated process would bring 100Kin10 partners together in small teams to assess problems and prototype potential solutions before final decisions (and funding) were made. To illustrate how radical this notion was, Saul generalizes how the grant process typically operates: "Organizations need to win RFPs to do their work, and then foundations will put an incentive in the form of some kind of grant, to say . . . we're looking for you to work collaboratively. You can [then] see each individual and their organization doing the mental math, to understand what hoop needs to be jumped through to get the monies . . . Am I 'collaborating' if I write someone else into the grant? Is that enough? Do I need more than one partner? Is there a way to do that collaboration so it doesn't interfere in our original design, plan, or conception of what we were going to do anyways?"

The traditional RFP process, which Saul and his colleague Sam Seidel are dramatizing for effect, assumes that some groups already have the answer; the task is just to find them and give them the resources (mostly funding) to apply their energies to good work. Though the best foundations allow for a dialogue between the grantor and grantee to discuss the goals together, many don't. In some cases, once the money is granted, the process effectively confines the successful grant winner to the activities specifically defined by the grant, so that any new or divergent needs discovered along the way must fit within the scope of the RFP. If Organization X finds out that its proposed Solution Y is simply not feasible, there's not a lot of room to come back and say, "You know, we got it wrong, and here's what we know now."

When Talia brought on the Business Innovation Factory as advisors, she was already convinced that a new approach was needed. As she explained, "Instead of running grants or even having an RFP, where your job as a grantee is to come in and say why you're better than everyone else, we decided to *flip the RFP*. We chose [a] challenge, and then invited people to all work together on that challenge. During the course work, they would talk to end users, they would design and prototype solutions. They would get feedback."

Lynda, as a grant recipient, says this process is very unusual. Despite being in the same field, educators typically have little opportunity to work with others outside their own organizations. The new approach changed that: "You get into a room with your peers, yet from very different walks of life, and share different takes on an issue . . . Instead of solving a problem, you're asked to spend some time with the issues, thinking and generating ideas of what *might* lead to concrete next steps. But the goal is not the transaction, the goal is to spend time thinking together. To ask, what has been successful, what has not? What do we think is needed? What are our assumptions and what do we not know that we might need to know?"

The premise, and promise, of collaborative work is that there's a space—what Saul calls "the gray area" and I call "the commons"—between all the familiar areas where everyone can meet to ask new questions. This is an area outside of what is already known to ask and then create what a group *needs* to know. When we spend time between our disciplines, between our existing understandings, we can, as Saul puts it, "combine and recombine in different ways to change the value equation or to solve a problem that's been difficult to solve before."

Elizabeth Rood, of the Bay Area Discovery Museum in San Jose, California, had been working on an idea with MIT to redesign an exhibit space into the world's first early childhood "fab lab," where kids could make things with their hands and, in doing so, learn the engineering necessary to build something. She heard that 100Kin10 was doing a fellowship program around engineering and thought, *What a cool way to think about [and fund] our fab lab!* She and her team were accepted into the program along with eleven other teams. While they knew the 100Kin10 focus was to explore ideas together, "We were as-suming whatever we ended up doing would be connected to our fab lab work. But we got into the first couple of days of the design thinking process and realized we had to let go of that. Not that we might not end up there, but we needed to not have that be our assumption."

Elizabeth explains that one challenge of "letting go" was knowing that organizations don't send you away from your day job to a confer-ence to "think together"; you're expected to argue for the benefit of your particular idea and get funding for it.

Another challenge was how messy the process actually was. "You have ideas, and then you get further along and your idea is less clear as you try to navigate with people you don't really know particularly well to come to different viewpoints. It's just a tricky process to get a group of strangers to do problem solving together." At one point, the groups took part in a puz-zle exercise, which Sam Seidel explains is a way for them to visualize the interdependencies among them, to notice that each person has something to give and something to get.* "You have lots of your own puzzle pieces, but not all of them are for you to solve your specific puzzle. None of us here have all our puzzle pieces, you have somebody else's and others have yours," is how the exercise was explained to the group. "And so, like a scav-enger hunt, you had to find someone who needed something you had."

But none of that helped as they engaged their real-life challenge.

After all, the puzzle pieces were tangible, and everyone agreed on their shape and who had which ones. The number of the puzzle pieces was known; they were even in the same room. Compare that to think-ing about solving the complexities of education and teacher training,

---

* Human relationships are informed by how we give and take from one another. Every time we interact with another person, we have a choice to make: Do we try to claim as much value as we can, or contribute value without worrying about what we receive in return? Adam Grant wrote about these tensions and trade-offs in his bestselling book *Give and Take* (New York: Viking, 2013). The 100Kin10 collaborative story showcases what Adam calls "successful givers," generous collaborators who don't ask for a transactional pay-back. They help because they have a shared purpose and they trust that in helping, they'll have an impact, and that their efforts won't be wasted.

where the pieces are abstract, with ownership and definitions all subject to interpretation.

The process bogged down; tension slowly built in the room.

Elizabeth says that at one point the facilitators—Sam and Talia—were
not being clear. "The murkiness of the process and the lack of clarity of
what we were supposed to be doing next was . . . really irritating."

It would seem that experts in education expect clarity in learning
how to learn.

Sam remembers this moment. "Yeah, we were collaborating in a
way that was new, and sort of codesigning it as we went. So during this
first round of the fellowship, in front of everyone, we were struggling.
And Talia hit the pause button, and the entire group would watch us
interact to say what's going on and work through it. We said what we
had planned to do was X, Y, and Z, but we were feeling like something
different was needed here and we're not sure we were right." This was
not, as it turns out, some theatrical mechanism to communicate a
skill. Talia and Sam were truly trying to figure out "what we were doing together, to shift course, together." They realized "it made other
people as uncomfortable as it did us."

And yet, this embodiment of the process of working through hard
stuff is important.

Just like the Ushahidi team worked together to figure things out,
what the 100Kin10 team was doing was asking everyone to figure
things out, too. When you're asking others to learn, exemplifying the
learning process yourself is a powerful contribution. Your own commitment to ongoing learning and growth eventually inspires the larger
effort, making room for others to risk making mistakes.* Good
intentions—"I want to learn"—are not enough. Learning how to learn
comes through practice, in the doing and faltering and then doing one
more time. Talia and Sam, intentionally or not, were helping cultivate

---

* The idea that your commitment infuses the larger effort is from Peter Senge, Hal Ham-
ilton, and John Kania's "The Dawn of System Leadership," *Stanford Social Innovation Review*, Winter 2015, http://ssir.org/articles/entry/the_dawn_of_system_leadership.

the conditions that let people see that failing in the process of learning together is okay. Needed, even.

It was also acceptable not to get it right at the outset. When some of the groups shared that they had done a great deal of thinking and yet did not "have an idea that works yet," they got "as much applause as those that had beautiful, fleshed-out prototypes," as Elizabeth recalls.

Gradually, Elizabeth's museum team did come to discover something of great significance to their own project: They realized that many of the target audiences that they were so eager to reach regarded engineering as being completely irrelevant to their lives. Engineers were still stereotypically viewed as guys with pocket protectors, so it would never have occurred to most of their target audience to visit the museum. No amount of repackaging of promotional campaigns for the children's fab lab exhibit would change that perception. Once they understood that, Elizabeth's team abandoned the fab lab idea and in its place conceived of an entirely new way to serve their community: a mobile lab that would travel to schools, decorated with inviting images that would highlight the accessibility of STEM-related activities. They would focus on things kids actually wanted to build and then show them how they were relevant to engineering and how engineering has value in everyday life.

As Elizabeth describes it, "This was a huge shift in institutional priorities. [It] happened very quickly, but I think we felt, because the three of us [the Bay Area Discovery Museum team included various people at the 100Kin10 workshop] and the larger group had worked through it all together by challenging all our assumptions, we knew this was the right thing to do. It was a better idea than our original idea, and we wouldn't have ended up there if we hadn't been willing to let go and rethink everything." The BADM contingent would ultimately share the idea with its team in San Jose, run a successful pilot, and get funding for this creative new solution.

Challenging your own assumptions, letting go of your existing agenda, and risking not coming up with anything useful are not easy.

Yet they are all central to getting large groups of people to cocreate bigger and better ideas.

This process is exactly how people commission themselves to solve new things.

And it matches the key shifts in leading complex change. In their book, *Leading from the Emerging Future*, Otto Scharmer and Katrin Kaufer describe three "openings" that help transform entire systems:

1. The opening of your heart lets you listen and be open to others shaping your thinking.
2. The opening of your mind is to be willing to examine and perhaps discard your deeply held assumptions.
3. The opening of the will enables the letting go of preset agendas.[16]

If you want to change the future, in other words, you have to also be willing to *be* changed. As Saul explains, "That's the hard work. You weed out a lot of people who are just trying to get a grant, trying to jump through hoops, until you're left with people who are willing to go into the process to figure out what they need to figure out." That means people who are willing to give and get, in a nontransactional way; people who trust in the process; and people who believe that the shared purpose is worth the risk of failure.

CREATE THE SAFE SPACE TO LEARN

In December 2013, the *New York Times* published a full-page editorial describing 100Kin10 as the "most important effort" in STEM teacher preparation in the country. In 2016, as part of National Teacher Appreciation Day, President Obama highlighted STEM educators who would not have been teaching had it not been for 100Kin10. Five years had passed since his call and that meeting at Ideo in New York City. The early group, which believed that getting ten thousand teachers prepared in two years was ambitious, now had 280 national partners, 35 of whom pledged a collective $90 million, and could count thirty thousand teachers trained to their credit. They were, also, *finally* on track to meet or exceed their goal.

What was it that led to these remarkable results?

As Talia explains, "Our little team"—100Kin10 is now a separate organization, spun off from Carnegie, with a staff of nine plus herself—"gets to punch above its weight class because of all the allies and colleagues who have become friends, and who have contributed and gone above and beyond." Talia praises the efforts of the community but doesn't like to talk specifically about what is working and why. "People have pushed us to share our wisdom. To opine. To put down a stake on a policy or a position. And for the most part we push back. That's because the value we have—the *real* value we offer—is creating a safe space for people to learn together . . .* [You] don't need me or us already knowing the answer; what you need us to do is to create the space so you can have your own answer."

To lead a group seeking to achieve a shared purpose is not to direct resources like a harbor master directing boats to port; instead, it is to create and maintain the conditions for people to figure out what they need to know, so they feel that ownership and responsibility to do it themselves.

Getting better at any skill is often described as a cumulative process. First you crawl, then you walk, then you run. Learning is usually understood as the path to perfection, to reaching the point where you know the answers. You proceed linearly until you reach the pinnacle. But in our lives many things are neither linear nor neat—not careers, or parenting, or markets. The actual process of creating something new is more like climbing a mountain that has no summit, where

---

* As a fan of organizational learning, I'll suggest multiple resources. One is Francesca Gino and Bradley Staats's piece, "Why Organizations Don't Learn," *Harvard Business Review*, November 2015, https://hbr.org/2015/11/why-organizations-dont-learn. Another is a timeless piece by Chris Argyris about double-loop learning that talks to the need to test your assumptions as a way to learn and solve tough things, like STEM education preparedness ("Double Loop Learning in Organizations," *Harvard Business Review*, September 1977, https://hbr.org/1977/09/double-loop-learning-in-organizations).

reaching one peak helps you understand the topography better so you can navigate to the next one.

GOING TOGETHER

Talia shares another of her favorite personal stories: "According to the *Legends of Our Fathers*, the Israelites were assembled at the edge of the Red Sea with the Egyptian army barreling down toward them, and the tribes of Israel were arguing amongst themselves, not able to agree on what to do. Then, Nachshon ben Aminadav took this step, a leap of faith, literally, and stepped in, and because of this act, the seas opened, and the Israelites walked together through to safety. People are hungry for this. For this call. They want to participate in something bigger, they want to take on this challenge, and they want to activate a part of themselves to step up and step in."

They want to go on the journey, together.

There are so many pressing problems in the world today: figuring out how to slow down climate change without collapsing economies, addressing underemployment as machines take on more work, resolving the tensions between antagonistic cultures, and of course, preparing future generations to deal with a world that requires more math and science skills. No country, organization, or highly gifted person knows how to solve any of those things independently.

But *we*, joined together in a shared purpose, could figure out a way to solve these critical issues—by bringing all that we have, by commissioning ourselves and one another, and then by starting that journey toward our shared vision and trusting that we can deal with not knowing long enough to actually learn new things. We need the leaders of ideas to believe that each of us *can* and *will* do amazing things if given the right conditions. We want our leaders to have faith in our ability to do the unforeseen, and to bring out the best sense of purpose in us all.

Then we can engage more of today's moon shots, as Talia and the members of her communities have so impressively done.

CHAPTER 9

―――――

Unlocking Onlyness

Design is not just what it looks like and feels like.
Design is how it works.
—STEVE JOBS

MAKING PERSONAL THAT WHICH IS SYSTEMIC

Up to now, we've focused on how *your* efforts unlock your ideas, even those ideas that others say are too wild.

In spite of existing power and status frameworks that say you can't do it, you will be a force for good, a change agent, and a dent maker. You will claim an idea, find allies with a common purpose, and doggedly organize together to make a dent, all because you, collectively, dream of a better way.

Proving that some large percentage of gaining power is simply you taking it.[1]

If we stopped here, though, the picture would be incomplete. Sometimes, maybe even oftentimes, we are inclined to see an issue as being entirely personal, putting all the onus on you, when it is *also* systemic. This is because it's hard to frame and explore systemic issues.

I'm as guilty of this as anyone.

When I first spoke about onlyness, at a 2012 Houston TEDx talk to a group that didn't know me, I focused my presentation on the "only" part of onlyness—the individual. The one indication that I offered the audience that there was more to the story was a single sentence: "Onlyness is fundamentally about honoring each person: *first* as we value ourselves, and *second* as we are valued." This was my way of pointing to the larger and fuller picture of what's involved—namely, the power of onlyness starts with your actions, *but* there's systemic work that existing power players can do to unlock onlyness.

If you're in a position to value the onlyness of others, then you need to know what follows. In this final section, we'll learn from Foldit and how they created a world-changing solution by unlocking the capacity of many, *systemically*.

FOLDIT: THE PATTERNS OF CHANGE

Until as recently as 2008, the scientific community had little understanding of the esoteric art of precisely how proteins fold. Though proteins are strings of amino acids, they don't spend much time in a linear, string-like form but rather exist as a three-dimensional shape. The process of transformation from string to shape is called "folding," and, while we could respond to this information by letting our eyes glaze over, it's important work. Misfolded proteins cause diseases such as Alzheimer's or ALS. The scientific community's lack of knowledge about how proteins fold or misfold has been a major impediment to finding cures to the diseases you and I might one day face.

Remarkably, despite technological advancements and a great deal of research, protein-folding patterns could not be reduced to a set of rules or software algorithms. As it turned out, discerning these patterns required human creativity and judgment, which meant, amazingly, humans could do what technology could not.

A kind of Manhattan Project to address the protein-folding issue was desperately needed, but a traditional approach of hiring experts to fold thousands of proteins would require finding and funding an army of experts at prohibitive cost. Instead, a team of researchers at the University of Washington set out to crowdsource the answer by creating an online game called Foldit.* The Foldit effort wasn't successful from day one, but when they failed, they asked new questions, ultimately designing a new system that created profound results. How they did so is highly instructive.

Adrien Treuille, an original member of the UW team, shares that from the beginning they saw Foldit as a way both to ask for help and to get highly skilled offshore labor. They initially envisioned that PhD students from around the world would lend their passion and expertise while enhancing their skills and even credentials, so the early iterations of the game were designed for just such players.

"But," Adrien learned, "requiring the 'players' to be alert to all the complexities of the science was a super-high bar for participation." Unfortunately, that meant the project had access to a relatively narrow pool of talent, all of whom had been trained in the same way, so that did not provide the necessary critical mass of fresh takes, creativity, and energy to enable Foldit to take off.

Adrien recalled how a meeting with some of the gurus in his field gave him an idea. "I was in a meeting with some of the world's best physicists and realized that instead of speaking in jargon-y language, they spoke more simply . . . We decided to do the same for Foldit. The theoretical stuff wasn't essential to engage creativity, we realized. What was *more* important was to let people learn concepts through trial and error so they could apply their own natural problem-solving skills. This got more people engaged, without lowering the quality of the results—at all."

Jargon is a shorthand, a kind of secret handshake that says you are either inside or outside our club. While jargon is not meant to filter out ideas, it often does.

---

* The Foldit team included David Baker, Zoran Popović, Seth Cooper, and Adrien Treuille.

So Foldit expanded the pool of potential players not by lowering the skills bar but by lowering the jargon bar. Changing the criteria opened the project to a broader pool of talent. New players could simply go to the Foldit website, create an account, and "play" some forty easily understood training games that demonstrated the logic of how proteins fold or misfold.

The participants were initially rewarded for their individual progress, but the participants themselves pointed out that this system led to much duplicated effort, because everyone was competing against each other instead of against the larger goal. The Foldit team responded by adding a point-based incentive scheme to reward collaboration; peers could give easy feedback for how people were helping one another. In general, the solo approach works best on problems if they are simple enough. But complex problems typically require combining different points of view, often across disciplines.

But those changes *still* weren't enough.

Adrien shares all this over coffee in a sun-dappled courtyard in Mountain View, California. It's clear that time has reduced the pain of what must surely have been deeply frustrating at the time. "Then we noticed," Treuille continued, "that the reward system that created collaboration *still* wasn't designed right. We realized that something was causing a distortion . . . that players tended to give more points to 'high-status' players—those that were identified with a top-tier school, for example, instead of those with the best, better results. So we tuned it again to hide those social status signals." When the factors that were not relevant—players' schools, the gender of the person, their age— were factored out, the emphasis turned to the ideas and the work itself.

The fine-tuning of the project in these three adjustments paid off.

A particular success came in 2011, when, in just ten days, volunteer Foldit players deciphered the structure of an AIDS-related virus that had puzzled scientists for fifteen years, a discovery that has helped the development of new medicines. In addition, a "protein folder's periodic table" database has been created that is accelerating research progress globally.

As the UW team wrapped up the work, Foldit held yet one more surprise. Who was the "winner" of the game, the best protein folder in the world? Was it the most credentialed expert? No. The most influential person in the field? No. A research student working at a top university? No. It was an individual who lacked authority, personal influence, and organizational heft. By traditional measures of power, it was someone who ought not to be "the best."

But from the perspective of onlyness, the "winner" makes perfect sense. The best protein folder turned out to be Susanne Halitzgy, an executive secretary at a rehab clinic in Manchester, England. Her day job was answering the phone, but her longtime passion was solving puzzles like Rubik's Cube and Sudoku. She had studied medicine early in her career but dropped it in part due to the sexism she encountered. Science had lost her capacity for a while, but Foldit's design unlocked it.

It's tempting to argue that Foldit had somehow enabled Susanne, but the fact is, she didn't change at all in her intelligence or her abilities. She did show up, of course, but how many of us have shown up and still not been counted? She did not become more broadly influential, and she did not experience any increase in her brand or reputation.

So what actually happened?

Her ideas, her contributions were *revealed*.

Foldit's structure and design was what enabled her crucial contributions alongside those of other players. By enabling a new set of conditions, Foldit made it possible for new value to be created for society. This is why unlocking onlyness matters: Benefit can accrue not only to a single party or a discrete group but also to society as a whole.

Foldit's creators not only designed something that worked better, they made something that opened up untapped capacity. Expanding upon the Steve Jobs quote that opens this chapter, you cannot consider *how* something works without asking *who* is it that it works for. And too often, not enough is done to ask how we can make what we're working on work for many.

Foldit's success shows us three systemic mechanisms that mattered. First, they had a generative approach to who should participate. Second, they structured their system to offer many ways to contribute to the outcome. And third, they rewarded the collective ingenuity of the group.

WHAT MAKES SOMEONE QUALIFIED?

Foldit's first systemic correction was to move away from credentials as a requirement because it wasn't a meaningful metric. This opened the project up to not only a *larger* set of players but a more *varied* one as well. It's hard to appreciate how big a shift that represented until you consider how most of the world judges who is qualified.

Most of us recognize that "highly skilled talent" is a key asset in our modern economy, because these individuals' ideas fuel growth and new products. This contemporary view of talent is such an improvement from the traditional practice of treating people as replaceable cogs in a machine that few question whether it, too, is still relevant.

We often evaluate people as being "qualified" in terms of specific check boxes, such as having the "right education," "right training," or "right experience," or even "knowing the right people." Although research has consistently shown that college entrance exams reveal more about socioeconomic background than innate capacity,* test scores are still used as a screening mechanism for top schools. Higher test scores then give you a leg up to open doors to the social and professional networks that form in those institutions, and to the companies that recruit from them. It's quite possible that the people who didn't score as highly could contribute just as much. Conflating education, or years of experience, with the capacity to contribute limits newness.

---

* Saul Geiser studied a sample of more than 1.1 million California residents who applied for admission to the University of California system between 1994 and 2011. In 1994, family income, education, and race together accounted for 25 percent of the variance in students' SAT scores, and parents' education was the strongest predictor. By 2011, the same socioeconomic background factors accounted for 35 percent of the variance in SAT scores, and race/ethnicity had become the most important factor. Saul Geiser, "The Growing Correlation Between Race and SAT Scores: New Findings from California," Center for Studies in Higher Education, October 27, 2015, http://www.cshe.berkeley .edu/news/growing-correlation-between-race-and-sat-scores-new-findings-california.

Foldit's is *such* a radically different framework that it's easy to skip past the question the team must have paused to ask itself. When not enough people are engaged, they could have assumed that there weren't enough qualified people, which would have assumed limited capacity. Instead, they challenged their definition of *qualified* and assumed there was an abundance of capacity, even if it was latent at the time. They thought critically about the criteria they were using to determine who could or could not participate and how to enable not only more but broader and unexpected participation.

Outdated criteria often remain as standards without anyone's analyzing why the criteria were established in the first place.

For example, many companies have a requirement that, to move up in a company, employees must have had a global assignment, or worked in operations. "Both of these criteria," says Marianne Cooper, a sociologist at the Clayman Institute for Gender Research, "can end up excluding women, because women are more likely than men to have partners that work, so this can mean they are less mobile. And women tend not to go into operations at the same rate as men. So, are these *really* the criteria that matter? If people really think about [it], and figure out what it is that really matters, they can revise things in such a way that with almost a snap of the fingers more people become 'eligible' to participate. It broadens the pool, and sets the stage for onlyness to emerge."[2]

The lesson is obvious: If we never question the underlying assumptions about who is qualified, we won't design to include the range of capabilities that onlyness enables.

DEFINE OUTCOMES MORE CAREFULLY

The Foldit team's second mechanism was structuring their system so that everyone could participate.

Most of us see ourselves as "fair" in terms of how we consider how people can contribute, without thinking of how we could be better. One particular idea in my field of technology is, *Hey, everyone has an equal chance because we're a meritocracy.* The word "meritocracy" always reminds me of a particular cartoon, one that is ubiquitous in the educational testing world even though its origin appears to be unknown.

In it, various critters—a bird, a monkey, a penguin, an elephant, a fish, a walrus, and a wolf—are lined up. A guy at a table in front of them says, "For a fair selection, everybody has to take the same exam: Please climb that tree."

The monkey is clearly ecstatic. The rest are dismayed.

Not a Meritocracy

That's because six of the seven, or 85 percent, will have no chance to contribute based on how the objective was defined. That's 85 percent of the creativity off the table. In real life, oftentimes, this is based on some criteria for "how things have always been done." Before the Foldit team lowered the jargon bar, they were needlessly blocking contribution in this way.

If Foldit hadn't adjusted to let all ideas have a shot, they might have ended up with the same solution they started with: *none.* But when you redesign the system so that many people apply their human ingenuity, then you are tapping into a wealth of capability. Like the water table below the surface of the earth, capacity is always there, but tapping it requires considered mechanisms.

REWARD SHARED OUTCOMES

Foldit's final key mechanism placed the onus of momentum onto the collective community. They said, "Solve it *together*," and then aligned the incentives to support that message.

Groups often get inconsistent messages. My team and I once worked with a big tech firm that had a labs division and wanted to take more ideas from the labs to market. It was a complex assignment, but the eventual answer was fairly simple. Predictably, the labs staff was rewarded for patentable work, but surprisingly, the rewards were greater for a solo effort than for a shared one, effectively incentivizing these brilliant people to work alone. Ideas consequently took years before they could be monetized, and because good products often require multiple patents, important collaboration was not taking place. The labs faced the possibility of closing entirely if this approach continued.

My team and I delivered our recommendation for change on a conference call with the head of the labs. He replied, "If we were smart, we would change"—but then went on to explain why change was impossible.

Sadly, his response is not atypical, as measurement, attention, and rewards are often directed toward individual performance. Managers regard this as the best way to achieve accountability: "How can we know *who* did *what* if we don't isolate the individual?" This mind-set is a holdover from the people-as-cogs era. When putting together a car on an assembly line, each person could be assigned a measurable task and even a minimum threshold —say, 123 widgets per day. Those who failed to meet their benchmarks could easily be identified and fired.

But in our modern economy, the emphasis on rewarding and isolating individuals—thinking IP and IQ are the end-all—defies logic.

Today, the context we work in together affects how we create value. The effect of collaborative culture is the exponential factor that raises the performance of the whole.[3]

Let's visualize Foldit's mechanisms in an unusual way: mathematically.

We begin by imagining a community of three participants, with abilities to contribute that we measure as, say, 2 and 5 and 10. If we can harness all that ability, we will get $2 + 5 + 10$, or 17. Now, let's expand the number of people involved, bringing in participants with abilities of 3 and 7 and 9. If the new members are diverse, there are better chances that their contributions are distinct (not duplicative).

If we're not careful, we might define a dynamic whereby only one member's contribution is valued, and perhaps not even the top contributor. We could end up with just 7 or 10, or even 2. But by defining a better, more generative outcome, we can benefit from all contributions. In our example, we can get the full sum, or 36—much better than even the best contributor.

So now we are accepting all contributions, and the participants are not yet fully collaborating. What if one member can only contribute 3 on her own, but excels at leveraging the gifts of others? Let's illustrate that with a multiplier effect. This could result in a potential $2 \times 3 \times 5 \times 7 \times 9 \times 10$, or 18,900. That's remarkable, and yet there are potentially more participants available. If we construct the system as the Foldit team did, where the system designers and the participants collaborate through a tug of mutual obligation to tune the system for better dynamics, we might view this as an exponential benefit, raising the contributions to a power such as a square: $18,900^2$ is over 357 million.

The numbers chosen above are playful, yet illustrative. And we can recognize that in the absence of Foldit's well-designed system, the contributions of many members would have been limited or lost entirely. How would we measure the total contribution in that case? How much more valuable was the Foldit system's overall contribution than zero?

The tug of mutual obligation to one another changes the performance of the whole. Much like Ushahidi, and the 100Kin10 team, you open yourselves up to identify breakthrough concepts. You all rise up to something higher, building off each other's ideas with fruitful arguments that make every idea better, pushing everyone to excel. Not only do you accomplish more than if you were working alone but, collectively, you achieve something much greater. When you ask people to come together around a common purpose, it invites them to bring everything they have. Then the measures and rewards—the ultimate tug of mutual accountability[4]—are what leads to astonishing results. It enables a group to achieve performance levels far greater than the individual bests of any team. This is what's possible, next.

How? By design.

Foldit's protein-folding game is certainly not the only system subject to inadvertent obstacles. We can see many examples in our larger economy: unpaid internships that make it possible only for those of upper-middle-class families to get much-needed work experience, or late-night meetings which limit the contributions of those who have family commitments but are where key relationships are built.

Do these systemic obstacles help us find new solutions? No, of course not.

The good news is that evidence is building. Research shows the "difference principle" can make a measurable impact, bringing increased revenues, more customers, greater market share, and greater relative profits.[*] The "difference principle" is also key to new solutions.[†] And it shows up in the bottom line; when organizations embrace a wide set of ideas, they have two to three times the performance results of companies that don't.[‡]

---

[*] Scott Page, at the University of Michigan, created a mathematical equation around the value of difference. Cedric Herring's research at the University of Chicago answers the question of how much diversity pays; see "Does Diversity Pay?: Race, Gender, and the Business Case for Diversity," *American Sociological Review* 74, no. 2 (April 2009): 208–24, http://journals.sagepub.com/doi/abs/10.1177/000312240907400203.

[†] Karim R. Lakhani, a Harvard-based open innovation professor, along with colleague Lars Bo Jeppesen, conducted a meta-analysis of 166 science challenges, involving 12,000 scientists, finding that "social marginality" played an important role in explaining individual success. Nearly 100 percent of winners came from "left field." See "Marginality and Problem Solving Effectiveness in Broadcast Search," *Organization Science* 20, accessed February 28, 2017, via Digital Access to Scholarship at Harvard, https://dash.harvard.edu/bitstream/handle/1/3351241/Jeppesen_Marginality.pdf?sequence=2.

[‡] For more on how collaborative models drive 2-3x firm performance, see Tim Kastelle, Nilofer Merchant, and Martie-Louise Verreynne, "What Creates Advantage in the 'Social Era'?"

But this is, even more importantly, an issue of prosperity. When onlyness is not enabled, our economy and our society are at risk. Today, nearly
60 percent of US workers (about 80 percent worldwide)* do work that
amounts to following orders. They are unable to contribute their ideas,
apply their creativity, or use their own judgment. This leaves broad parts
of our workforce underutilized, a situation that will become only more
pronounced with the automation of many jobs. The ability to contribute
and do good work is so basic a need for human beings that when there is
no good work to be had, they can easily turn resentful and bitter.

This results in not only a failing economy but a failing society.

> **In a world** where fitting in has been rewarded more than
> standing out,
>> it has hobbled our own ideas.

> **In a world** where differences can preclude getting counted,
>> it limits growth and innovation.

> **In a world** of automation and jobs being replaced by machines,
>> we are losing the sense of dignity that comes
>> with contributing what each of us can.

Notice that the implication is not about how we feel or that there's
inclusivity of all people—important ideas, surely—but that the systemic frameworks affect ideas, growth, and dignity, and so can limit
our own personal and global prosperity.

We each want to add value and be valued. We need to design systems to enable that.

Innovations 10, no. 3/4 (2015): 81–91, http://www.mitpressjournals.org/doi/pdf/10.1162/inov
_a_00241.

*   These facts are based on the work of Richard Florida and the Martin Prosperity Institute,
whose work intersects social implications of economic trends. Richard is currently a professor at the Rotman School of Management at the University of Toronto, and is a leading
thinker about how creative work blossoms. His research shows that the global number
of people doing creative work—work that requires independent judgment, decision making, and idea generation—is 23 percent. That leaves 77 percent doing work that can be
easily replaced by automation.

CHAPTER 10

Powerful Beyond Measure

You may say I'm a dreamer, but I'm not the only one.
—JOHN LENNON

WHO ARE YOU?

Who are you? The answer to this question shapes everything, from *how you create value* to *how you are valued.*

Now, as short as that question is, it's not an especially easy one to answer. As I can personally attest: I remember about six years ago, when I was at some professional event, someone turned to me, putting their hand out to shake mine, and asked the question, "And who are you?" Not an unusual question at a professional event, but I looked around at the circle of eight or so people now staring at me and my mind went blank, and I replied,

"I am nobody."

I was as surprised as anyone to hear those words come out of my mouth. Even as they spilled out, I wanted to rope them back in.

Yet I *had* said it. There was a reason why. I normally would have answered the question "Who are you?" by sharing my title or my organization's name. Right around the time of this party, I had been CEO of a professional consulting firm that had served just about every major tech company in Silicon Valley. But, because of deeply personal circumstances, I had just decided to wind down my firm and was feeling quite confused about my role, my value, and even my future when I answered the question that particular night. A few colleagues happened to be in that circle, and they jumped in, adding information about my background and résumé, but I never forgot how displaced I felt without that job to hang my hat on.

Maybe you can remember a time when you felt displaced, and you have a parallel story in which you felt small when you responded to the question, "Who are you?"

It haunted me, and so I started to wonder why I responded as I did.

The experience exemplifies how power has worked.

When I said "I am nobody," I made myself powerless by my own characterization. I could have easily used alternative descriptions—for example, that I was recently retired from leading a strategy firm, or that I was looking for how next to apply my passions for driving innovation processes, or that I was ready to think about things broader than business. Of course, there's a reason why I didn't do that. In my response, I had thoroughly embraced and embodied the social constructs of power as they exist today. External things like jobs, credentials, and so on *are* power constructs that shape and constrain who gets a seat at the table, who gets to count. By buying into them, I simply reinforced them.

People should never define themselves—or what they can or cannot do—in any way that limits them. You are not nobody; you are powerful beyond measure.

This is why onlyness matters.

It says that the keys to power are in your hands—but you must use them.

THE ARC OF HISTORY

During the two years I spent dedicated to researching and writing this book, I lived with my family in Paris, which added texture to my perspective on the power of onlyness. From our apartment on Rue de Bellechasse, Normandy was a two-hour train ride away and the Louvre a fifteen-minute walk, literally something I walked by almost daily. While most people view them as tourist checklist items, these landmarks tell a larger story about power, about doors that are open or barricaded shut. Normandy was the launching point of many conquests, and the Louvre was built in the twelfth century as a fortress under King Philip II.

These are physical reminders that for much of human history, the fates of "civilized people" were largely controlled by a single person, a royal, who would send armies into battle to die for ego or wealth. Only his ideas counted, and he dictated who got what, who was deemed valuable. If anyone else's "wild" ideas counted, it was because they got permission from the king. In the thirteenth century, some twenty-five powerful barons were able to take a seat at the table from King John of England, a narrow but momentous shift toward distributed power codified in the Magna Carta. Later still, when the United States won its independence, suffrage in the new democracy was comparatively broad compared to that in Britain. It included *regular* people. Of course, we see today that they were all white, male, and property holders, but still, it was far more inclusive than ever before, at about 6 percent of the population.

The Magna Carta folks and the people of the United States didn't politely ask for power, they demanded it. In both of those examples, people banded together to insist. Bit by bit since then, the fraction of humanity that has been able to control its own destiny has grown— not because they were given permission but, instead, because they insisted on their ability to guide their own future and acted accordingly.

The topic of power is not considered a polite topic, so it is often avoided. But it is central to our lives, because it affects the destiny of our lives, of our ideas. It governs who gets to create value and how they are valued. That's worth fighting for.

This is the arc of history playing itself out. More people count, and so do their ideas.

And that's the promise of onlyness.

RADICAL?

You may not be surprised to learn that some have told me that the concept of onlyness is too far-fetched, too revolutionary, and even radical. They say this while trying not to sound disapproving, but it's clear that they're slightly embarrassed to know me. Their tone implies that I should be wary of promoting such things.

How do I think of this question of onlyness as a radical notion? To answer that, I return to the etymology of the word "radical," from the Latin *radicalis*, meaning "of or having roots." I believe that at the very root of our humanity is a passion to create value with heart, to work alongside others who care, and to make a difference. I believe that each of us has something of value to offer—all 7.5 billion of us. While not everyone *will*, anyone *can*. The fact that today so many people do not is not a sign that they lack capacity, but instead it's a sign that the scaffolding and structures need to be built to let them do so. This is society's problem, and its opportunity.

So, if onlyness *is* radical, then I hope it is also capable of changing the root of our economics and our societies to realign them with our fundamental nature as humans. Economics, my own field of study, is most fundamentally concerned with understanding the rules and power dynamics that enable people to best create and capture value. The word "value" here is an interesting one, because it is both an economic term and a human one. Onlyness can unlock both. Onlyness is that rich, yet largely untapped capacity in our economy and our society.*

---

* Here, I am drawing from the capabilities approach of Amartya Sen, the Nobel Prize–winning economist who believes that measuring capacity is a key consideration if we are

So yes, onlyness is radical. As history has shown us, though, what was once considered *radical* is now understood as a fundamental truth.

WHAT'S NEXT

When I first started asking the question, *Is onlyness a new way in for even wild ideas?* I didn't really know what I would find.

One part of me optimistically believed that I would discover something useful, something that would honor the might of a wild idea, that would enable quite possibly *anyone* to contribute. How profound that could be! A lever of change that could shatter the status quo and pull all of us into the future.

What might happen if the power of onlyness were broadly
     embraced by the world?
Who might contribute?
What could they create?
What old problems might finally be solved through a fresh
     perspective?
What new opportunities would open up?
And more to the point: What new dents could be made?

But, to temper that optimism, it's important to note that to make it a reality will take action. Yours. Mine. Ours. For us to create a world where onlyness is celebrated, we're going to have to live it.

We're going to have to give ourselves permission to have an original idea, even when no one else is advocating it. We're going to have to signal our passions and seek our allies. We're going to have to honor

---

to fix our larger problems. He argues that if you measure only GDP, you are assessing the economic output of a nation by its physical products. But that measures only one factor that matters. Measuring capacity evaluates something more important: the *ability* to create value, which may or may not have a dollar sum attached to it or have a physical product to show for it. See Thomas Wells, "Sen's Capability Approach," *Internet Encyclopedia of Philosophy*, http://www.iep.utm.edu/sen-cap.

the onlyness of all the people we meet along the way, even if, and especially when, we disagree with them. We're going to have to reframe the questions people pose and change the conversation to find new ways through. We're going to have to learn how to lean on one another so that we can build trust and scale our ideas. We're going to have to learn how to galvanize those who might not experience what we know to be true by showing them its value. We're going to have to engage people to work with us, not by telling them the answers but by pointing toward a new horizon.

And, more than anything else, we're going to have to give these ideas free rein to charge ahead so they can pull us into the future. All progress is born of new ideas. They let us reimagine who we are and how we might be. Ideas rupture the status quo and incubate the future.

A future that works for not just the few but for many.

When I first started writing this book, I would wake up in the middle of the night scribbling out notes about who I hoped would use it—a list of people who wanted to make a dent: "For the person who wants healthy behavior to be an everyday choice, for the person who wants to create equality for her daughters, for those who want all kids to learn the fun of science, for those who want to make their cities green, for the person who wants to make it easy to get locally grown foods, for the person who wants to stop political correctness, for the person who wants to make sure no baby dies because of lack of food or prenatal care, for the person who wants to cure AIDS, for the person who wants to house the homeless," and so on. Sometimes the items were lofty, and sometimes low to the ground. I would toss one night's list out the following morning, only to compose a new one the next week.

The book I did wind up writing is, quite simply, for you. You, who have suppressed your ideas, struggled to find "your people," or been told your vision for what was possible was simply "too much." For you, who knew deep inside that you had a lot to offer the world, if you could just find a way to give it.

I write in the hope that big changes like what I experienced at community college are no longer lucky one-offs but a way for so much more to happen. Whatever your particular dent is, I want you to see yourself as a shaper of destiny, a constructive builder of community and galvanizer of growth. For those who are responsible for existing systems, I want you to see that you can redesign what we do and how we do it, so that we enable the full power of onlyness. You will dent the world to include the best ideas. And, together, dent by dent, we'll reshape the world.

I hope if someone asks you, *Are you powerful enough to make a dent?* you'll answer, *Yes*. And, you will add, *Let's*. Let's go dent the world.

ACKNOWLEDGMENTS

How do I name even a single person and not the hundreds of others who contributed in meaningful ways to what you are holding in your hands?

Do I mention the publishing team whose near-term birthing was crucial without mentioning the thousands of people in audiences—most of whose names I could never know—who heard me beta test my thinking aloud?

How far back do I go? From 2010, when I answered the question of *Who are you* with *I am nobody*? Because that's when I started asking myself a series of new questions about the power of one's own worth and ideas as a relationship of social constructs, though I wouldn't have described it that way at the time. It was really hundreds of exchanges, with the many who nudged, coached, and challenged me that led to this clarity.

Is it a matter of weighing some people's ideas as more important than others? Ugh, no. After all of this celebration of onlyness, let's not revert to hierarchy.

Let me, rather, use this space to express several meta-acknowledgments, and then say my individual *merci* to everyone, one by one, with T-shirts or flowers, and even handwritten thank-you notes that your grandma would be proud of. For now, here come the metas, and please understand that many people do not fit neatly into just one, single category.

First, I am grateful to all those whose very lives showed me the reality of onlyness, for of course the book truly began from them. I'm grateful to everyone who helped me dig deeper into so many stories, both those few that appear in this book and those hundreds that I was not able to include. Thank you to all the people who helped me, in a hundred different ways, conceive of how to articulate the notion of onlyness, from questions raised at my various global keynotes to the book's chapter structure, and on down to the specific ideas and words. I also must acknowledge the many who have helped me with the complex logistical work of turning a lot of words, and a few pieces of art, into a book, and then getting that book into the hands of readers. Lastly, I'm especially grateful to my family and closest friends for their tolerance and resilient support during the part roller-coaster ride, part jungle trek that is the journey of probably any book, but especially so for this one. You have all helped manifest this idea of onlyness.

Thank you, all.

Thank you, thank you, thank you.

Nilofer

NOTES

CHAPTER 1: Arriving at the Question

1. The blog he started, in 2006, was called *Neophyte Humanitarian*. It can be viewed at http://neophytehuman.blogspot.com.
2. The website, called WinWinWiki, was absorbed by Appropedia; see http://www.appropedia.org/WinWinWiki.
3. Measuring something like this precisely is difficult. But Lonny Grafman, one of Appropedia's founding team who now teaches a graduate course on development technology, had hands-on experience for ten years when he observed, *Half of the development projects I see in the field are spending their time solving problems that have already been solved somewhere else, sometimes as near as the next village over.* Interviewee Lonny Grafman, in discussion with the author, February 28, 2017.

CHAPTER 2: Starts with You

1. Tokenism is not the only form of power that *appears* to create equal participation levels but in *actuality* does not. Sherry R. Arnstein's article "A Ladder of Citizen Participation," *Journal of the American Planning Association* 35, no. 4 (1969): 216–24 is an extremely useful framework. The article can be read at http://www.participatorymethods.org/sites/participatorymethods.org/files/Arnstein%20ladder%201969.pdf.
2. Patrick Thibodeau, "IT Jobs Will Grow 22% Through 2020, Says U.S.," *Computerworld*, March 29, 2012, http://www.computerworld

.com/article/2502348/it-management/it-jobs-will-grow-22--through -2020--says-u-s-.html.

3. "Meet San Jose's Toyota Standing O-Vation Recipient," Oprah.com video, 6:40, http://www.oprah.com/oprahstour/the-toyota -standing-o-vation-recipient-in-san-jose-video. Nyatche Martha's name was provided by BGC staff.

4. Kim Zetter, "Nicholas Christakis: Does This Social Network Make Me Look Fat?" *Wired*, February 12, 2010, https://www.wired.com /2010/02/ted-2010-nicholas-christakis-does-this-social-network -make-me-look-fat.

5. Hannah Braime, "What Japanese Pottery Can Teach Us About Feeling Flawed," Becoming Who You Are, April 29, 2014, http:// www.becomingwhoyouare.net/japanese-pottery-can-teach-us -feeling-flawed/.

6. John W. Gardner, speech delivered at the Hawaii Executive Conference, Kona, HI, April 1993.

7. Tom Peters, "The Brand Called You," *Fast Company*, August 31, 1997, https://www.fastcompany.com/28905/brand-called-you.

8. On power and status having reinforcing loops, see Joe C. Magee and Adam D. Galinsky, "Social Hierarchy: The Self-Reinforcing Nature of Power and Status," *Academy of Management Annals* 2, no. 1 (2008): 351–98.

9. "Trail to Eagle," National Eagle Scout Association, accessed February 28, 2017, http://www.nesa.org/trail.html.

10. "Zach Wahls Speaks about Family," YouTube video, February 1, 2011, https://www.youtube.com/watch?v=FSQQK2Vuf9Q.

11. W. W., "The Iowa House v Zach Wahls and His Moms," *Economist*, February 4, 2011, http://www.economist.com/blogs/democracyin america/2011/02/politics_and_morality_gay_marriage.

12. Andrew Solomon's book *Far from the Tree* (New York: Scribner, 2012) traces how identity is shaped.

13. Michael Martinez, Amanda Watts, and Deanna Hackney, "Gay Scout's Request for Eagle Rank Rejected," CNN, January 9, 2013, http://www.cnn.com/2013/01/08/us/california -gay-eagle-scout.

14. Karen Andresen, "Overturn Ban on Gay Scouts," Change.org, accessed February 13, 2017, https://www.change.org/p/overturn -ban-on-gay-scouts.

15. Ibid.

16. Interview with Zach Wahls, March 12, 2015, in New York, NY.

CHAPTER 3: Discovering Yours

1. James G. March, *Explorations in Organizations* (Palo Alto, CA: Stanford University Press, 2008).

2. Interview with Herminia Ibarra, September 16, 2014, at Laterie, Paris, France.

3. Cheryl Strayed, *Brave Enough* (New York: Knopf, 2015).

4. Santa Clara University Management Department chair Terri Griffith, conversation with the author, 2015.

5. Charlie Guo, "How I Crashed and Burned in YCombinator," *Backchannel*, January 2, 2015, https://backchannel.com /how-i-f-ed-up-in-ycombinator-35a19e7ace68#.w4ukkzr6m.

6. For more on this topic, see John Henry Clippinger, *A Crowd of One: The Future of Individual Identity* (New York: Public Affairs, 2007).

7. Cara Buckley, "Spreading the Word (and Pictures) on 'Real' Sex," *New York Times*, September 7, 2012, http://www.nytimes.com/2012 /09/09/fashion/cindy-gallops-online-effort-to-promote-real-not -porn-fed-sex.html.

8. Elizabeth Gilbert, *Big Magic: Creative Living Beyond Fear* (New York: Riverhead, 2015).

9. Nilofer Merchant, "Be Very Afraid for a While," *Yes and Know* (blog), November 15, 2011, accessed February 28, 2017, http://nilofermerchant .com/2011/11/15/be-very-afraid-for-a-while.

CHAPTER 4: Find Your People

1. Melissa Stanger, "The 17 Coolest Co-Working Spaces in America," *Business Insider*, January 8, 2013, http://www.businessinsider.com /the-17-coolest-coworking-spaces-in-america-2012-12?op=1/#dy-hall-9.

2. Anil Dash, "Nobody Famous," AnilDash.com, May 24, 2015, accessed February 21, 2017, http://anildash.com/2015/05/nobody -famous.html.

3. Sherry Turkle, *Alone Together* (New York: Basic Books, 2011).

4. Doree Shafrir, "Tweet Tweet Boom Boom," *New York*, April 26, 2010, http://nymag.com/news/media/65494/index6.html.

5. Derek Sivers, "How to Start a Movement," TED talk, 3:09, February 2010, https://www.ted.com/talks/derek_sivers_how_to_start_a _movement.

6. Michael Arrington, "Too Few Women in Tech? Stop Blaming the Men," TechCrunch, August 28, 2010, https://techcrunch.com/2010 /08/28/women-in-tech-stop-blaming-me.

7. Rachel Sklar, "Flashback: 'Women in Tech' Panel, TechCrunch Disrupt, Sept. 2010," Medium.com, March 13, 2014, accessed February 21, 2017, https://medium.com/thelist/transcript-women -in-tech-panel-techcrunch-disrupt-sept-2010-51af6ef51e27.

CHAPTER 5: Common Purpose, Not Commonalities

1. Mitch Wagner, "Death Threats Force Designer to Cancel ETech Conference Appearance," *Information Week*, March 26, 2007; Heather Havenstein, "Death Threats Force Woman to Suspend Blog, Cancel Talk at O'Reilly Conference," *Computerworld*, March 27, 2007.

2. "Blog Death Threats Spark Debate," *BBC News*, March 27, 2007.

3. Clay Shirky, "A Group Is Its Own Worst Enemy," edited keynote speech, O'Reilly Emerging Technology Conference, April 2003, http://www.shirky.com/writings/herecomeseverybody/group _enemy.html.

4. Amanda Lenhart, "Cyberbullying," Pew Research Center, June 27, 2007, http://www.pewinternet.org/2007/06/27/cyberbullying/.

5. Justin Jouvenal, "Stalkers Use Online Ads as Weapon Against Victims," *Washington Post*, July 14, 2013, http://www.washingtonpost .com/local/i-live-in-fear-of-anyone-coming-to-my-door/2013 /07/14/26c11442-e359-11e2-aef3-339619eab080_story.html.

6. http://topics.nytimes.com/top/reference/timestopics/people/c/tyler _clementi/index.html.

7. Jon Ronson, "How One Stupid Tweet Blew Up Justine Sacco's Life," *New York Times Magazine*, February 12, 2015, https://www.nytimes .com/2015/02/15/magazine/how-one-stupid-tweet-ruined-justine -saccos-life.html.

8. The advertising-based business models are rewarding this outcome. While mob behavior destroys lives on the Internet, the Internet companies themselves are raking in monies. Jon Ronson, in his 2015 book *So You've Been Publicly Shamed*, sized Google's income for the Justine Sacco incident. He worked with economists and online advertising experts to document that Google alone earned between $120,000 to $456,000 in just one month for just this singular incident.

9. "The Darkest Corner of the Internet," Anti-Semitism.net, accessed February 21, 2017, http://www.anti-semitism.net/southern-poverty -law-center/the-darkest-corner-of-the-internet-how-reddit-became-a -haven-for-vile-hate-speech.php.

10. Jason Pontin, "Such 'activists' are not principled, neutral defenders of free speech or civility. They are proponents of group and class interests." June 11, 2015, 11:20 a.m. Tweet, https://twitter.com/jason _pontin/status/609017391222300672.

11. Anil Dash, "If Your Website's Full of Assholes, It's Your Fault," AnilDash.com, July 20, 2011, accessed March 14, 2017, http://anildash .com/2011/07/if-your-websites-full-of-assholes-its-your-fault.html.

12. *Situation Analysis of Women and Children in Pakistan* (Islamabad, Pakistan: United Nations Children's Fund, 2012), http://www.unicef .org/pakistan/National_Report.pdf.

13. Gillian Felix, "Every Three Seconds, a Girl Is Traded as a Swara," Plain Talk BM, July 22, 2013, http://www.plaintalkbm.com/swara.

14. Humberto Maturana and Francisco Varela, *The Tree of Knowledge: The Biological Basis of Human Understanding*, revised ed. (Boston: Shambhala Publications, 1992).

15. Tehmina Ahmed, "Lambs to the Slaughter," review of *Swara*, directed by Samar Minallah, *Newsline*, September 2003, http://www .newslinemagazine.com/2003/09/lambs-to-the-slaughter.

16. Samar Minallah, "Judiciary as a Catalyst for Social Change," public litigation plea, Pakistan Supreme Court, 2003, http://www.supreme court.gov.pk/ijc/Articles/9/2.pdf.

17. Zeba Ali, "Cover Up, or Else . . . ," *Newsline*, May 2009, http://newslinemagazine.com/magazine/cover-up-or-else.

CHAPTER 6: Without Trust, You Don't Scale

1. Eddie Huang, interview by Joe Rogan, *Joe Rogan Experience*, YouTube video, 14:17, February 28, 2013, https://www.youtube.com/watch?v=_hwLMBdnbXk.

2. Wesley Yang, "Eddie Huang Against the World," *New York Times Magazine*, February 3, 2015, http://www.nytimes.com/2015/02/08/magazine/eddie-huang-against-the-world.html.

3. Tom Rielly, "William Kamkwamba and Me," *Tom Rielly Blog* (blog), June 6, 2014, accessed February 24, 2017, http://tomrielly.typepad.com/trielly/2014/06/william-kamkwamba-and-me-our-seven-year-adventure.html.

4. Adam Kahane, *Power and Love* (San Francisco: Berrett-Koehler, 2010).

5. "Exposed: Andrew Wakefield and the MMR-Autism Fraud," BrianDeer.com, accessed February 24, 2017, http://briandeer.com/mmr/lancet-summary.htm.

6. James Law, "University of Sydney Study Rules Out Link Between Vaccination and Autism," NewsComAu, May 20, 2014, http://www.news.com.au/lifestyle/health/university-of-sydney-study-rules-out-link-between-vaccination-and-autism/story-fneuz9ev-1226923177732.

7. "Public Trust in Government: 1958–2014," Pew Research Center, November 13, 2014, http://www.people-press.org/2014/11/13/public-trust-in-government.

8. Aja Romano, "How the Alt Right's Sexism Lures Men into White Supremacy," *Vox*, December 14, 2016, http://www.vox.com/culture/2016/12/14/13576192/alt-right-sexism-recruitment.

9. James West, "97 Percent of Climate Scientists Can't Be Wrong," *Mother Jones*, May 16, 2013, http://www.motherjones.com/blue-marble/2013/05/video-97-climate-scientists-cant-be-wrong.

10. Jonathan Weiner, "Curing the Incurable," *New Yorker*, February 7, 2000, http://www.newyorker.com/magazine/2000/02/07/curing-the-incurable.

11. David Ruth and Jeff Falk, "Rice U Study: Managers Can Boost Creativity by 'Empowering Leadership' and Earning Employees' Trust," press release, Rice University Office of Public Affairs, October 8, 2014, http://news.rice.edu/2014/10/08/rice-u-study-managers-can-boost-creativity-by-empowering-leadership-and-earning-employees-trust.

CHAPTER 7: Galvanizing Many to Care (Enough to Act)

1. "Oral History: Leo Bretholz," Holocaust Encyclopedia, United States Holocaust Museum, video, 4:14, accessed February 20, 2017, https://www.ushmm.org/wlc/en/media_oi.php?ModuleId=0&MediaId=2953.

2. "Holocaust Survivors Seek Apology, Full Accounting from SNCF," *Maryland Daily Record*, video, 2:55, March 3, 2011, https://vimeo.com/20627352.

3. *Freund v. SNCF*, docket no. 09-0318 (2d Cir. 2010), http://caselaw.findlaw.com/us-2nd-circuit/1228873.html.

4. Ariel Schwartz, "A $53 Billion Plan to Bring High-Speed Rail to the U.S.," *Fast Company*, February 8, 2011, http://www.fastcompany.com/1725228/53-billion-plan-bring-high-speed-rail-us.

5. Katherine Shaver, "Opposition to Maryland Rail Line Bidder Raises Questions About Accountability for Holocaust," *Washington Post*, March 9, 2014, https://www.washingtonpost.com/local/trafficandcommuting/opposition-to-maryland-rail-line-bidder-raises-questions-about-accountability-for-holocaust/2014/03/09/ddb2a8c2-9f0c-11e3-a050-dc3322a94fa7_story.html.

6. Leo Bretholz and Rosette Goldstein, "SNCF: Pay Reparations to Victims of the Holocaust," Change.org, accessed February 28, 2017, https://www.change.org/p/sncf-pay-reparations-to-victims-of-the-holocaust.

7. Change.org started in 2007 with a broader set of cause-related tools. *Stanford Social Innovation Review* has published a case study in the

evolution of the organization: Greg Beato, "From Petitions to Decisions," Fall 2014, https://ssir.org/articles/entry/from_petitions _to_decisions.

8. Leo Bretholz and Rosette Goldstein, "SNCF: Pay Reparations to Victims of the Holocaust," Change.org petition, accessed February 28, 2017, https://www.change.org/p/sncf-pay-reparations-to-victims -of-the-holocaust.

9. Ibid.

10. Lawrence Mishel and Alyssa Davis, "Top CEOs Make 300 Times More Than Typical Workers," Economic Policy Institute, June 21, 2015, http://www.epi.org/publication/top-ceos-make-300-times -more-than-workers-pay-growth-surpasses-market-gains-and-the -rest-of-the-0-1-percent.

11. "Are They Worth It?" *Economist*, May 8, 2012, http://www.economist .com/blogs/graphicdetail/2012/05/ratio-ceo-worker-compensation.

12. Mattathias Schwartz, "Pre-Occupied," *New Yorker*, November 28, 2011, http://www.newyorker.com/magazine/2011/11/28/pre-occupied.

13. Michael Levitin, "The Triumph of Occupy Wall Street," *Atlantic*, June 10, 2015, http://www.theatlantic.com/politics/archive/2015/06 /the-triumph-of-occupy-wall-street/395408/.

14. Tom Cohen, "5 Years Later, Here's How the Tea Party Changed Politics," CNN.com, February 28, 2014, http://www.cnn.com/2014 /02/27/politics/tea-party-greatest-hits.

15. Levitin, "The Triumph of Occupy Wall Street."

16. "Morehouse President John Wilson Interviewed by Hank Williams at Platform Summit 2013," YouTube video, 34:02, September 22, 2014, https://www.youtube.com/watch?v=aoToxS66IKY.

17. Jennings Brown, "One Year After Michael Brown: How a Hashtag Changed Social Protest," *Vocativ*, August 7, 2015, http://www .vocativ.com/218365/michael-brown-and-black-lives-matter.

18. Jelani Cobb, "The Matter of Black Lives," *New Yorker*, March 14, 2016, http://www.newyorker.com/magazine/2016/03/14/where-is -black-lives-matter-headed.

19. Judith Ohikuare, "Meet the Women Who Created #Black LivesMatter," *Cosmopolitan*, October 17, 2015, http://www

.cosmopolitan.com/entertainment/a47842/the-women-behind
-blacklivesmatter.

20. Alex Altman, "Person of the Year: The Short List: No. 4: Black Lives
Matter," *Time*, February 28, 2015, http://time.com/time-person-of
-the-year-2015-runner-up-black-lives-matter/.

21. Deen Freelon, Charlton D. McIlwain, and Meredith D. Clark, *Beyond
the Hashtags: #Ferguson, #Blacklivesmatter, and the Online Struggle for
Offline Justice* (Washington, DC: Center for Media and Social Impact,
February 2016), http://cmsimpact.org/resource/beyond-hashtags
-ferguson-blacklivesmatter-online-struggle-offline-justice.

22. Jon Swaine, Oliver Laughland, Jamiles Lartey, and Ciara McCarthy,
"Young Black Men Killed by US Police at Highest Rate in Year of 1,134
Deaths," *Guardian*, December 31, 2015, https://www.theguardian
.com/us-news/2015/dec/31/the-counted-police-killings-2015-young
-black-men.

23. Wesley Lowery, "Aren't More White People Than Black People
Killed by Police? Yes, but No," *Washington Post*, July 11, 2016, https://
www.washingtonpost.com/news/post-nation/wp/2016/07/11
/arent-more-white-people-than-black-people-killed-by-police-yes
-but-no/.

24. Ibid.

25. You can learn more about Dr. Schradie's work on her website:
http://schradie.com.

26. Jon Caramanica, Wesley Morris, and Jenna Wortham, "Beyoncé in
'Formation': Entertainer, Activist, Both?" *New York Times*, February
6, 2016, https://www.nytimes.com/2016/02/07/arts/music/beyonce
-formation-super-bowl-video.html.

27. "About Six-in-Ten Americans Say More Changes Needed to Achieve
Racial Equality," Pew Research Center, June 21, 2016, http://www
.pewsocialtrends.org/2016/06/27/3-discrimination-and-racial
-inequality.

28. "(1857) Frederick Douglass, 'If There Is No Struggle, There Is No
Progress,'" Blackpast.org, accessed February 28, 2017, http://www
.blackpast.org/1857-frederick-douglass-if-there-no-struggle-there-no
-progress.

29. Julie Hirschfeld Davis and Michael D. Shear, "Unrest over Race Is Testing Obama's Legacy," *New York Times*, December 8, 2014, http://www.nytimes.com/2014/12/09/us/politics/unrest-over-race-is-testing-obamas-legacy-.html.

30. Michael D. Shear and Liam Stack, "Obama Says Movements Like Black Lives Matter 'Can't Just Keep on Yelling,'" *New York Times*, April 23, 2016, http://www.nytimes.com/2016/04/24/us/obama-says-movements-like-black-lives-matter-cant-just-keep-on-yelling.html.

31. *Investigation of the Ferguson Police Department* (Washington, DC: United States Department of Justice Civil Rights Division, March 4, 2015), https://www.justice.gov/sites/default/files/opa/press-releases/attachments/2015/03/04/ferguson_police_department_report.pdf.

32. "Justice Department Opens Pattern or Practice Investigation into the Baltimore Police Department," press release, United States Department of Justice, Office of Public Affairs, May 8, 2015, https://www.justice.gov/opa/pr/justice-department-opens-pattern-or-practice-investigation-baltimore-police-department.

33. John Eligon and Sheryl Gay Stolberg, "Baltimore After Freddie Gray: The 'Mind-Set Has Changed,'" *New York Times*, April 12, 2016, http://www.nytimes.com/2016/04/13/us/baltimore-freddie-gray.html.

34. Brittany Packnett, "I Sat Beside Obama at the Black Lives Matter Meeting. This Was No Political Show," *Guardian*, February 20, 2016, http://www.theguardian.com/commentisfree/2016/feb/20/barack-obama-black-lives-matter-meeting.

CHAPTER 8: Commission to Own It

1. Patrick Meier, "How Crisis Mapping Saved Lives in Haiti," *National Geographic*, July 2, 2012, http://voices.nationalgeographic.com/2012/07/02/crisis-mapping-haiti.

2. Ryan Ozimek, "Snowmaggedon of Interest in Our Crowdsourcing Site," *Picnet* (blog), February 12, 2012, http://blog.picnet.net/tag/snowmageddon/.

3. Team Ushahidi Flickr album, https://www.flickr.com/photos/ushahidi/albums.

4. Erik Hersman, "The Trust Bridge," Ushahidi.com, May 5, 2011, https://www.ushahidi.com/blog/2011/05/05/the-trust-bridge.

5. Amy Gallo, "Making Your Strategy Work on the Frontline," *Harvard Business Review*, June 24, 2010, https://hbr.org/2010/06/making-your-strategy-work-on-t.html.

6. Part of a jacket quote for Robert H. Frank's *Success and Luck: Good Fortune and the Myth of Meritocracy* (Princeton, NJ: Princeton University Press, 2016).

7. Ann O'Neill, "What Went Wrong with Jahar? Answers Elude Boston Bomber's Defense," CNN.com, May 9, 2015, http://edition.cnn.com/2015/05/08/us/tsarnaev-13th-juror-penalty-defense.

8. "2013 Boston Marathon Attacks: Please Upload Any Photos in Relation to the Attacks That You Have," Reddit.com, April 15, 2013, accessed February 28, 2017, http://www.reddit.com/r/boston/comments/1cf5wp/2013_boston_marathon_attacks_please_upload_any.

9. Source for the two blameless men and where it was published came from Salon research: Laura Miller, "The Boston Marathon and Reddit: When the Internet's Deluded Amateur-Hour Detectives Ran Amok," *Salon*, April 15, 2015, http://www.salon.com/2015/04/15/the_boston_marathon_and_reddit_when_the_internets_deluded_amateur_hour_detectives_ran_amok.

10. Boston Police Dept., "CAPTURED!!! The hunt is over. The search is done. The terror is over. And justice has won. Suspect in custody." April 19, 2013, 8:58 p.m. Tweet, https://twitter.com/bostonpolice/status/325413032110989313.

11. "Reflections on the Recent Boston Crisis," Redditblog.com, April 22, 2013, http://www.redditblog.com/2013/04/reflections-on-recent-boston-crisis.html.

12. "Blow Minds. Teach STEM," accessed February 28, 2017, http://blowmindsteachstem.com.

13. "JFK Moon Speech," YouTube video, 7:56, uploaded by newcurioshop, February 28, 2009, https://www.youtube.com/watch?v=iEu45U0zRZ4.

14. Interview with Lynda Kennedy, May 16, 2016.

15. Interview with Saul Kaplan and Sam Seidel of the Business Innovation Factory, May 5, 2016.

16. Otto Scharmer and Katrin Kaufer, *Leading from the Emerging Future* (San Francisco, CA: Berrett-Koehler, 2013).

CHAPTER 9: Unlocking Onlyness

1. Jeffrey Pfeffer, "80% of Power Is Taking It," *Talent Quarterly*, https://www.talent-quarterly.com/80-percent-of-power-is-taking-it.

2. From an e-mail exchange between the author and Marianne Cooper on August 21, 2016.

3. Anita Williams Woolley et al., "Evidence for a Collective Intelligence Factor in the Performance of Human Groups," *Science*, October 29, 2010, http://science.sciencemag.org/content/330/6004/686.

4. Jon R. Katzenback and Douglas K. Smith, "The Discipline of Teams," *Harvard Business Review*, July/August 2005, https://hbr.org/2005/07/the-discipline-of-teams.

INDEX